D1432244

Ion-Exchange Sorption and Preparative Chromatography of Biologically Active Molecules

MACROMOLECULAR COMPOUNDS

Series Editor: **M. M. Koton,** *Institute of Macromolecular Compounds*
Leningrad, USSR

ION-EXCHANGE SORPTION AND PREPARATIVE CHROMATOGRAPHY OF
BIOLOGICALLY ACTIVE MOLECULES
G. V. Samsonov

Ion-Exchange Sorption and Preparative Chromatography of Biologically Active Molecules

G. V. Samsonov
Institute of Macromolecular Compounds
Leningrad, USSR

Translated from Russian by
R. N. Hainsworth

Translation Edited by
D. C. Sherrington
University of Strathclyde
Glasgow, Scotland

CONSULTANTS BUREAU • NEW YORK AND LONDON

Library of Congress Cataloging in Publication Data

Ion-exchange sorption and preparative chromatography of biologically active molecules.

(Macromolecular compounds)
Bibliography: p.
Includes index.
1. Ion exchange chromatography. 2. Biomolecules – Analysis. I. Samsonov, G. V. (Grigoriï Valentinovich) II. Series.
QP519.9.I54I66 1986   543′.0893   86-19644
ISBN 0-306-10988-3

This translation is published under an agreement with the Copyright Agency of the USSR (VAAP).

© 1986 Consultants Bureau, New York
A Division of Plenum Publishing Corporation
233 Spring Street, New York, N.Y. 10013

Printed in the United States of America

Foreword

This book deals with the physico-chemical principles underlying ion-exchange sorption and chromatography. It does not cover in any detail the experimental and instrumental aspects of practical separations. The author has developed the subject starting from the synthesis and structure of the ion exchangers employed, through the thermodynamics of sorption selectivity and the equilibrium dynamics of ion sorption, to the kinetics and dynamics of non-equilibrium ion-exchange systems.

Throughout this treatment the additional factors arising from the exchange of complex organic ions, as opposed to simple mineral ones, have been interwoven. The author has stressed the application in the separation of organic ions with biological activity, many of which are synthesized in biotechnological processes, and in view of this he uses the expression "physico-chemical biotechnology." In practice, however, his in-depth treatment is applicable to any charged organic species with multifunctionality and/or high molecular weight, and is therefore by no means restricted to biologically active materials, and certainly not to those molecules from a biotechnological source. Bearing this in mind, the text has a much wider value than the title may convey.

Throughout his work the author has used the term "ionite" routinely as a very general expression for any ion exchanger. Most often these are typical ion-exchange resins based on crosslinked vinyl monomers, or poly-condensation matrices. However, the term also encompasses inorganic-based ion-exchange materials, and occasionally linear polyelectrolytes used as models for the network systems. The derivatives "cationite" and "anionite" therefore have the same relationship as "cation exchanger" and "anion exchanger."

The text and accompanying figures and tables have been considerably streamlined by the use of abbreviations which have been gathered together for convenient reference. In this context the term "sulfocationite" has been coined to indicate a cation exchanger carrying a sulfonic acid group.

D. Sherrington

Preface

The presentation of ideas adopted in this monograph is largely
the result of many years of research and teaching at the Institute of
Macromolecular Compounds of the USSR Academy of Sciences, Leningrad.
I have given a course at the Chemical Pharmacology Institute, Leningrad,
for 25 years called "The Theoretical Basis of Fine Physico-chemical
Biotechnology" to final year students in the Faculty of Technology.

I consider myself fortunate that I can deal with both fundamental
and applied problems at the same time when studying systems that include
synthetic polymers (especially electrolytes) and organic and physiologically active substances (especially ions), and carry out scientific and
applied research on the separation, purification, and fractionation of
antibiotics, enzymes, hormones, amino acids, nucleotides, and many other
groups of organic substances. Several groups of workers in different topic
areas are united by their use of the general principles and concepts set
out in this monograph, which covers ion exchange and preparative ion-
exchange chromatography using mixtures of organic and physiologically
active substances.

The name of the course "The Theoretical Basis of Fine Physico-chemical
Biotechnology," a name suggested over 25 years ago, is in keeping with the
theoretical analysis of the scaling-up of processes for obtaining physiologically active substances under conditions in which they are chemically
unreactive. This includes the processes of purification, isolation, and
separation, as well as the methods of forming derivative structures to
stabilize and modify the physiologically active substances to give them new
properties. These may be to prolong the substances' activity or to direct
them to specific targets in the organism. In this monograph we shall
consider only some of these topics and shall analyze ion-exchange sorption
and the preparative chromatography of physiologically active substances.
Over the last 25–30 years the area has been turned into a separate division
of physical chemistry so that we can now use a small number of initial
ideas to predict and establish the best materials, regimes, and methods for
carrying out a preparation.

The last few years have also seen a tumultuous development of biology
and biotechnology that has led to the appearance of a series of names for
separate disciplines that can all be said to come within the science of
universal chemical biology. Thus it can be maintained that areas such as
bioorganic chemistry, biotechnology (the technology of directed biosynthesis), bioengineering, and technological enzymology all fall into this
category. However, I believe it is right to keep to the term "fine
chemical biotechnology" since the area of physico-chemical biotechnology
described here lies within the definition given above.

In the treatment of the concepts of fine physico-chemical biotechnology priority has been given to those that have appeared as a result of the work I or those under my supervision have carried out.

This book is a treatment of the theoretical basis of modern large-scale separation, isolation, and purification of physiologically active substances using network polyelectrolytes and dynamic column regimes. Practical problems must be overcome on the basis of theoretical analyses of the physical chemistry of network polymer electrolytes, their thermodynamics and kinetics, and the dynamics of intermolecular interactions between them and organic ions. In this analysis, I draw mainly from the theoretical and experimental work of my colleagues and myself. The methods that have been developed have quite a general nature and can be applied to a variety of sorptive and chromatographic systems that lie on the fringes of the investigations covered here.

The first chapter justifies the idea that large-scale efficient methods for separating organic electrolytes can be created by combining the selectivity of the interactions of network polyelectrolytes and the sharp boundary regimes of dynamic frontal processes. That the separating power of frontal processes can be raised to that of contemporary eluant processes is demonstrated. This is done mainly by raising the quantity of substance to be separated to a value tens of thousands of times higher than that for a preparative eluant process in a comparably sized installation.

The second chapter considers the permeability of network polyelectrolytes. The morphology of net structures in the swollen state may be appraised only by using a number of investigative methods. The need for synthesizing heteronet structures (a class including macronets) that would be osmotically stable to changes in the ionic strength and pH of the environment, in order to improve the equilibrium and kinetic permeability of the polyelectrolyte and the reversibility of its sorption of complex ions, is emphasized. The thermodynamic and kinetic mobility of elements of the net structure is discussed. This effect was first demonstrated at the Institute of Macromolecular Compounds in Leningrad.

The third chapter is the largest and contains a thermodynamic analysis of intermolecular interactions in systems involving network polyelectrolytes and organic counter-ions. The interactions of weak and strong polyelectrolytes with weak and strong electrolytes in the surrounding solution are all considered from the same point of view. The concept of polyfunctional interactions between organic counter-ions and ionites is generalized and the cooperative, statistical, and concentration effects are considered. Quasi-ideal and non-ideal ionite systems are covered theoretically and experimentally. The large quantity of experimental evidence given in the chapter supports the validity of the principles derived.

In the fourth chapter I analyze the equilibrium dynamics of ion exchange based on concepts I have developed. Starting with column (dynamic) processes involving ions with the same charge, I pass on to more complicated systems that involve ions with different charges, weak electrolytes and weak polyelectrolytes, and those that involve two and three phases. Criteria are introduced which characterize the formation of sharp-boundary ion zones when the flow rate is low, and which can be used to predict the regimes needed for efficient preparative chromatography. Experimental data are provided that demonstrate the usefulness of the criteria.

The fifth chapter justifies new ways of intensifying separation processes for physiologically active substances that diffuse slowly in network polyelectrolyte beads. The theoretical analysis of the kinetics and dynamics of ion exchange is important when fast efficient methods are to be developed. What are needed are large-capacity sorbents that have good hydrodynamic properties without requiring expensive high-pressure equipment. The problem is overcome by using regimes that lie on the boundaries between quasi-equilibrium and regular processes, and by using new ionites for which the diffusion path for organic ions is short (e.g., surface-layer or bidispersed ionites). The theoretical analysis provided dimensionless numbers which can be used to describe the yield from quasi-equilibrium regimes. The new ionites and the theory have been used to realize the efficient large-scale liquid chromatography of physiologically active substances at high solution flow rates and short completion times. A series of examples at the end of the chapter illustrate separation and fractionation with these conditions.

I would like to thank my colleagues for their suggestions and comments. In preparing this manuscript I received the assistance of many collaborators and I would most like to acknowledge my debt to them all.

G. V. Samsonov

Contents

CHAPTER FOUR
EQUILIBRIUM DYNAMICS OF ION SORPTION AND
STANDARD QUASI-EQUILIBRIUM FRONTAL CHROMATOGRAPHY

CHAPTER FIVE
KINETICS AND NON-EQUILIBRIUM ION-EXCHANGE DYNAMICS

Abbreviations

The following are expansions of the abbreviations used in the text. Where there is not an obvious or meaningful English "translation," the original Russian abbreviation is transliterated. Abbreviations marked with an asterisk are trademark designations and so are transliterated. When an abbreviation is suffixed, the suffix indicates differences in crosslink ratio or synthesis conditions. Sulfo is a prefix indicating that a matrix is derivatized with a sulfonic acid group.

Sulfocationites

KU-2*	Sulfonated copolymer of styrene and divinylbenzene
KRS*	Sulfonated copolymer of styrene and p-divinylbenzene
KU-6*	Sulfoacenaphthene condensed with formaldehyde
KU-5*	Sulfonaphthalene condensed with formaldehyde
KU-1*	Sulfophenol condensed with formaldehyde
KFS	Phenoxyethylsulfonic acid condensed with formaldehyde
KFSS	β-Phenoxyethanesulfosalicylic acid condensed with formaldehyde
KFSN	β-Phenoxyethanesulfonic acid condensed with β-naphthol and formaldehyde
KNS	β-(Naphthoxy)ethanesulfonic acid condensed with formaldehyde
KU-23*	Macroporous sulfonated copolymer of styrene and divinylbenzene
SNK-E	The sodium salt of 2-N-methacryloylamino-8-naphthol-6-sulfonic acid copolymerized with N,N'-ethylenebismethacrylamide
SNK-H	The above salt copolymerized with N,N'-hexamethylene-bismethacrylamide
SNK-D	The same with divinylbenzene
SNK-M	The same with methylenebisacrylamide
SDV	Sulfonated copolymer of styrene and divinylbenzene
SBS	Sulfonated copolymer of styrene and butadiene
SKB	Phenoxyethylsulfonic acid condensed with phenoxyacetic acid and formaldehyde
SFMAHMD	Copolymer of 4-sulfophenylmethacrylamide and N,N'-hexamethylene-bismethacrylamide
SK	Surface-layer cationites which are impermeable glass spheres covered with a sorbing layer of a styrene/divinylbenzene copolymer

Carboxylic Cationites

CPA	Phenoxyacetic acid condensed with formaldehyde
CPAC	Phenoxyacetic acid condensed with chlorophenol and formaldehyde
KRFU	Phenoxyacetic acid condensed with resorcinol and formaldehyde
KRFFU	Phenoxyacetic acid condensed with resorcinol, phenol, and formaldehyde
KB-4P-2*	Copolymer of methacrylic acid and divinylbenzene
KB-4*	Copolymer of methacrylic acid and divinylbenzene

KB-2	Copolymer of acrylic acid and divinylbenzene
KMDM	Copolymer of methacrylic acid and N,N'-alkylenebismethacrylamides
KADM*	Copolymer of acrylic acid and N,N'-alkylenebismethacrylamides
Biocarb-T*	Copolymer of methacrylic acid and 1,3,5-triacryloyltriazine
KM-2P*	Saponified copolymer of acrylonitrile and divinylbenzene
Biocarb-S*	Copolymer of methacrylic acid, p-aminosalicylic acid, and 1,3,5-triacryloyltriazine
SLCI	Surface-layer carboxylic ionite. Glass core and sorbing layer of copolymer of methacrylic acid and N,N'-hexamethylene-bismethacrylamide
SGK-7*	Copolymer of acrylic acid and divinylbenzene
MK-30*	Copolymer of methacrylic acid, ethylene glycol monomethacrylate, and ethylene glycol dimethacrylate

Anionites

ASD*	Aminated chloromethylated copolymer of styrene and divinyl-benzene
ASB*	Aminated chloromethylated copolymer of styrene and butadiene
ARA*	Aminated chloromethylated copolymer of styrene and p-divinyl-benzene
FAF	Phenoxyethyltrimethylammonium salt condensed with formaldehyde
FAKh	Phenoxyethyltrimethylammonium salt condensed with p-chlorophenol and formaldehyde
AV-16	Condensation product of pyridine and polyethylene-polyamines with epichlorohydrin
AV-17	Aminated chloromethylated copolymer of styrene and divinyl-benzene

Crosslinking Agents

DVB	Divinylbenzene
DV	Divinyl
B	Butadiene
HTA	1,3,5-Triacryloyltriazine
EDMA	N,N'-Ethylenebismethacrylamide
HMDMA	N,N'-Hexamethylenebismethacrylamide
DMDMA	N,N'-Decamethylenebismethacrylamide

Miscellaneous Abbreviations

MAA	Methacrylic acid
AKA	Acrylic acid
MANS	2-(N-Methacryloyl)-amino-8-naphthol-6-sulfonic acid
PMAA	Polymethacrylic acid
PAA	Polyacrylic acid
PVSA	Polyvinylsulfonic acid
T	Telogen (usually carbon tetrachloride)

Polycondensation	Condensation polymerization

1
Introduction

1.1. THE PROBLEMS OF FINE PHYSICO-CHEMICAL BIOTECHNOLOGY

The methods of biotechnology, such as directed microbiological synthesis, biochemical methods of breaking down complicated organic compounds
(especially those using the biological catalysts - enzymes - and immunobiological enzymes in particular), and the methods of fine organic synthesis have recently been very successfully applied to obtaining physiologically active substances (PAS's). The substances that have been synthesized and biosynthesized include proteins (enzymes), polypeptides and
amino acids, nucleic acids, nucleotides, nucleosides, polysaccharides, low-
molecular-weight hydrocarbons, and a variety of others that have a physio-
logical effect, such as antibiotics, regulators of various types, hormones,
and vitamins. Both natural and synthesized PAS's are at first complex
mixtures of substances, and the initial solutions may contain tens, hundreds, or even thousands of components, many of which are closely related
to the desired product in terms of their physico-chemical properties. If
the products of a chemical synthesis are very specific (the intermediate
and side products being only admixtures), then during a biosynthesis a
multicomponent solution will be formed. The culture liquid obtained is a
very complicated mixture from which we have to isolate only one, or a small
number of, components. Natural products too are no less important and
include primary extracts from animal and vegetable tissues and solutions
that have been obtained at the first stage of separating out a natural PAS.
In all these, the separation of an individual or very pure PAS is a complex
problem that can be solved using modern and effective physico-chemical
methods. This then is the subject area of physico-chemical biotechnology.

The sources for a PAS are shown in Figure 1.1. Natural substances
were historically the first to be used, even though their purity was
improved and their structure established only gradually. The organic
synthesis of PAS's, if we discount preparation of chemically simple rep-
resentatives, has been particularly well developed in the last decade.
Modern methods of fine organic synthesis enable us to carry out such complicated multistep operations as the synthesis of the hormone insulin, or
oligonucleotide and polypeptide synthesis. The feature of these methods is
their high single-step yields of the basic products, and these are due to
the application of the principles of modern organic chemistry in the area
of reactivity - specific catalysis and the blocking of groups which cannot
then participate in the reaction. Many of these processes require dozens

1

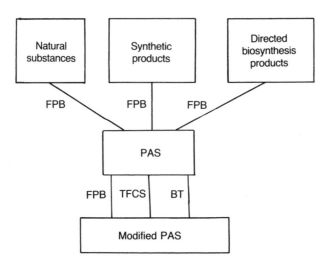

Fig. 1.1. Technology and preparative methods of obtaining physiologically active substances. TFCS means technology of fine chemical synthesis. BT means biotechnology (directed synthesis, genetic engineering, and biochemical enzymatic technology). FPB means fine physico-chemical biotechnology.

of steps. Carrying out this sort of organic synthesis on a preparative or industrial scale is included in Figure 1.1 under the heading of the technology of fine chemical synthesis. A great deal of attention has been paid in recent years to biosynthetic methods. These use microbes to obtain the products and apply concepts from molecular biology, molecular genetics, and genetic engineering. Strains are created that can produce a given PAS, e.g., an antibiotic, in quantities hundreds or thousands of times larger than can natural micro-organisms. The only way of improving on the biosynthesis process itself is to make a rational choice of the medium and aeration conditions and to introduce components that are the precursors of the PAS's. Microbiological synthesis can be used to obtain antibiotics, vitamins, enzymes, and some other PAS's. Modern biotechnology does not only embrace direct microbiological synthesis, but also includes biochemical processes that use biocatalyzing enzymes, particularly the isomerizing and immuno-enzymes. Modifying PAS's have recently received considerable attention.

The products of biosynthesis may be transformed into more active substances using the methods of chemical synthesis. Apart from the greater variety of preparations, this approach enables us to obtain, for example, antibiotics to which pathogenic micro-organisms cannot develop resistance. The products of chemical synthesis may also be transformed into new, even more active substances using a biosynthesis or the methods of biotechnology.

It would be senseless simply to obtain a PAS without being able to separate, isolate, and purify it using the methods of fine physico-chemical biotechnology. These methods are in fact used to obtain PAS's from natural sources. Deriving biochemicals and medicines from animal and vegetable raw material requires extraction, precipitation, crystallization, ion exchange, and chromatography, and until recently derivations such as that of insulin from animal pancreases and alkaloids from plants required dozens of stages and could go on for weeks. Using the methods of fine physico-chemical biotechnology (FPB) to get pure substances from the products of microbe synthesis is no less important. Any attempt to play down this task and

to regard it simply as another "chemical purification" and then use traditional procedures will result in only low yields of the pure preparation
that does not contain any possibly toxic side-products close in properties
to the main product. The formulation of a general theory for the preparative separation of complex mixtures of substances and the development of
modern methods of FPB has enabled both the purity to be improved and the
preparative methods and technological procedures to be made less expensive.
The methods of FPB, which include the sorptive and chromatographic methods
we described here, have been given much attention in the last few years for
preparing both medicines and biochemicals for analysis and research. These
methods are less important when synthetic PAS's are being prepared than
when natural PAS's are being isolated or when microbiological methods are
being used, but even here they are essential for attaining the final aim of
getting a pure or individual substance efficiently. The modification of a
PAS from an organic synthesis using a biosynthesis or vice versa will also
require FPB methods at several stages. A special area where these methods
are used is the physico-chemical modification of a PAS, for example,
sorptive immobilization or complex formation in solution. A PAS modified
in this way has not been chemically transformed but has acquired new
properties that enable it to be stabilized and to be given new features, in
particular to be transformed using macromolecular carriers in preparations
with long activity or using substances capable of localizing its effect to
particular organs.

1.2. PHYSICO-CHEMICAL METHODS OF SEPARATING MIXTURES

There are two chief ways of separating a mixture. One group of
methods is based on interphase mass exchange, and these are often used to
establish an equilibrium or quasi-equilibrium in a heterogeneous system.
A difference in the partition coefficients of the substances between the
different phases is the main factor used to separate them. The second
group of methods includes techniques in which the separation is due to
differences in the rates at which different substances migrate in a force
field. These kinetic methods are usually implemented in a single, though
perhaps not homogeneous, phase. It should be noted that there are a few
techniques that use both principles, i.e., a heterogeneous system and
kinetic separation. In these methods the rate at which the substances go
from one phase to the other defines the rate at which the components are
separated. An effective interphase partition is achieved due to the
kinetic factors common to many processes, particularly the dynamic ones.
What is most important is that the kinetic factors enhance the effect of
the equilibrium interphase transfer, or at least do not hinder the separation. In those processes (such as dynamic and chromatographic ones) in
which the kinetics of the interphase transfer can worsen the separation,
the effects of the kinetic factors have to be lessened.

The main techniques for separating mixtures in heterogeneous systems
are shown in Table 1.1. The most widely used kinetic methods are shown in
Table 1.2. The methods in Table 1.1 in which the components may be found
in the gas phase cannot be used for physiologically active substances
because most of them are destroyed in the gas phase. However, evaporation
and condensation, especially fractional distillation, are widely used as
auxiliary technological methods to produce PAS's. Gas-liquid chromatography is exceedingly important for analyzing PAS's and their associated
substances, and it is possible to use modifications of PAS's that will
vaporize into the gas phase.

Until now the most widely used preparative methods for separating and
purifying PAS's have mostly been single-interaction processes that use a
liquid and a solid phase, or two liquid phases. Separating substances to

3

Table 1.1. Single-Interaction and Multi-Interaction Methods for
Separating Substances in Heterogeneous Systems

Phases	Single-interaction	Multi-interaction
G/L	Evaporation and condensation	Fractional distillation
	Dissolution of gases	Gas-liquid chromatography
G/S	Molecular adsorption	Gas chromatography
L/L	Extraction	Counter-current extraction Liquid partition chromatography
	Membrane methods	Cascaded membrane systems
L/S	Precipitation and crystallization	Precipitation chromatography
	Molecular adsorption	Molecular liquid chromatography
	Ion exchange	Ion-exchange chromatography
	Fusion	Zone refining
	Biospecific sorption	Affinity chromatography

Table 1.2. Kinetic Methods for Separating Mixtures

Method	Phases
Diffusion	Liquid (diffusion through membranes in the gas and liquid phases)
Sedimentation	Liquid-solid
Electrophoresis	Liquid-gel

analyze them requires many steps and most often chromatographic methods are
used. Until the 1970's the only good analytical technique was gas-liquid
chromatography, which could separate and analyze, after several minutes or
tens of minutes, a mixture whose components' properties were very close. Liquid
chromatography, e.g., an amino acid analysis on an ion-exchange resin (ionite),
was much less efficient because of the slow rate of achieving interphase
equilibrium and the kinetic spreading of the component zones. Our modern
techniques came into being after the whole process (and particularly the
sorbents and apparatus) had been developed so that micron sorbent beads
could be used in high-pressure equipment[1-5]. Modern high-efficiency
high-pressure liquid chromatography is not inferior to gas-liquid chromatography in terms of separating power and speed of operation, since there
has been considerable success using analytical liquid chromatography of
macromolecular substances, particularly proteins. Progress has been made
in improving the efficiency of gel-permeation chromatography (gel chromatography) of proteins and other biological macromolecules, while the chromatography of synthetic polymers is approaching that of low-molecular-
weight substances.

Gel-permeation, or exclusion, chromatography cannot usefully be
analyzed as a single-interaction process. In the classical version of this

4

method there is either no interaction between the porous column substrate and the absorbed materials or it is insignificant. However, we should point out that it is possible to use the restricted permeability of sorbent particles and to apply molecular or ionite sieves for preparative purposes. These, in contrast to normal gel-chromatography packings, have high energies of interaction between the substances and sorbent beads. When there is a high absorptive capacity, it then becomes possible to use single-interaction processes to separate small molecules, or their ions, from larger ones which cannot penetrate into the sorbent bead. Later we shall demonstrate that dynamic methods based on molecular and ionite sieves are essentially single-interaction processes.

While considering the separation methods for macromolecular compounds it should be noted that the success of gel chromatography has reduced the significance of sedimentation chromatography, which at one time attracted attention due to the difficulties of implementing the sorptive chromatography of macromolecules. When appraising affinity chromatography we should bear in mind that it is specific to only one component (though in rare cases to several), and so analytical techniques based on it for proteins and enzymes are, from this view point, limited. The success of affinity chromatography is due to the creation of biospecific sorbents that can bind onto substances in a solution using enzyme-substrate, enzyme-inhibitor, or antigen-antibody interactions. The suppression of the sorptive properties of biospecific sorbents (the deactivation of their active centers) is often observed when isolating substances from extracts or culture liquids without initial purification. It is often difficult to scale up these processes and use biospecific sorbents repeatedly, especially when working with multicomponent solutions. The main success of affinity chromatography has been in the laboratory isolation of proteins (enzymes included) after several pretreatment stages.

Laboratory preparation of physiologically active substances is carried out using countercurrent extraction and sometimes zone refining.

Multi-interaction dynamic processes are thus basic for the analysis and small-scale separation of mixtures. The efficiency of elution chromatography is very high, and interphase exchange is repeated many times in the columns. The height of a theoretical plate for the high-performance liquid chromatography of a PAS is close to a millimeter and sometimes even less. Thus modern analytical columns can have thousands of theoretical plates. Naturally such repeated interphase transfer corresponds to a much more effective separation than can be achieved in a single-interaction process. However, the cost of multi-interaction processes, particularly chromatography, limits its large-scale use. The usual ratio of separable mixture to sorbent in pure elution chromatography is between 10^{-5} and 10^{-6}, while increasing this ratio impairs the degree of separation. At the same time, the diameter of the column is also limited. Increasing the diameter increases the spreading of the substance zones. This is due to uneven penetration and dissimilar sorption between the center and periphery (especially at the walls) of the column. Elution chromatography, which is mostly employed as an analytical technique, is also used preparatively with a ratio between substance and eluant of 10^{-3} to 10^{-4}. The high plant costs, particularly for high pressures, remains a negative factor. Using elution chromatography on a large scale in the laboratory or even on an industrial scale is only tolerable for very costly products that are used in small quantities, e.g., vitamin B_{12}. Despite the lure of the high efficiencies of multi-interaction processes and the desirability of using them as industrial processes, single-interaction processes remain the more important. The principal techniques for treating PAS's industrially are extraction, ion exchange, precipitation, crystallization, molecular adsorption, and membrane processes.

Most of the diseconomies of scale we have mentioned here with regard to multi-interaction processes also apply to kinetic processes (Table 1.2). Thus, they are relevant only for analytical or small-scale preparative preparations.

While we are considering the large-scale preparative isolation of physiologically active substances, particularly using ion exchange, we should emphasize that the creation and use of highly specific sorbents and the modern theory of the equilibrium, kinetics, and dynamics of ion exchange have opened up new possibilities for separating, isolating, and purifying PAS's.

1.3. PREPARATIVE SELECTIVE SORPTION AND CHROMATOGRAPHY

If we are to develop successful methods for separating large quantities of PAS using liquid chromatography at larger-scale laboratory, pilot plant, and industrial levels, we must start from the principles that encompass both the efficiency and economics of a process. The use of high pressure, and the attendant high-cost plant, for separating substances that are not themselves valuable or applied in tiny quantities is usually impracticable for preparative, let alone industrial, chromatography. The more economic alternatives are the single-interaction sorptive methods. The selective sorption of a single substance from a solution containing many components must use highly specific sorbents. Recently, considerable success has been achieved in synthesizing very specific sorbents that can selectively absorb particular groups of organic substances. We shall discuss this topic in a number of sections later.

The direct use of selective sorbents in the form of dispersed suspensions that are then separated from the solution is not convenient operationally and technologically. Selective sorption by passing a solution through a column is not only technologically better, it is also more efficient. Completely saturating the column is usually a single-interaction process in which the sorbed substance comes into equilibrium with the original solvated state. However, the absorptive capacity of the sorbent in a dynamic regime is considerably greater than that in a static one because of the possibility of forming sharp boundaries between the substances being sorbed in the column, and this leads to the establishment of an equilibrium with the original solution. Figure 1.2 shows graphs of sorptive capacity and equilibrium concentration for static and dynamic sorptive processes under conditions for the formation of sharp fronts. The case given in Figure 1.2 is true not only for molecular sorption but also for ion-exchange sorption.

Preparative large-scale chromatography includes both selective sorption in the column and selective desorption. A successful isolation of one particular substance from a multicomponent solution is based on a frontal selective sorption-desorption process under conditions that allow the formation of sharp boundaries between the zones of the substances being separated. An understanding of the equilibrium dynamics of sorption was basic for the development of many of the preparative methods of isolating antibiotics, hormones, alkaloids, enzymes, and many other PAS's[6-16]. Usually difficulties arose in implementing the selective desorption under conditions that would allow sharp boundaries to form between the substance zones. The introduction of equilibrium dynamics criteria[14,16] allowed workers to predict and implement regimes for fully displacing the substances being separated by ion-exchange chromatography. The regimes are formulated by choosing the constants of the selective desorption as the competing components are varied by using the solution concentrations and by selecting the solvents and pH.

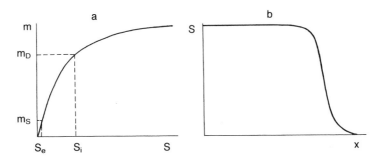

Fig. 1.2. Static and dynamic sorption capacity. a) Sorption isotherm.
b) Distribution of the concentration of the sorbate in the
column. S is the concentration of the sorbate in the column, m
is the equilibrium sorption capacity, x is the distance from the
outermost layer of sorbent in the column, S_i is the initial
concentration of sorbate in the column, S_e is the equilibrium
concentration in the solution for static sorption, m_D is the
dynamic sorption capacity, and m_S is the static sorption
capacity.

The hypothesis that the efficiency of frontal processes for isolating
substances from complicated mixtures is less than that of elution chromatography is not always well founded. The exceedingly high separating
capacity of elution liquid chromatography is estimated from columns that
contain thousands of theoretical plates per meter. Naturally, a virtually
single-interaction frontal process should not be described in terms of this
parameter. However, elution chromatography must operate with a small
(close to unity) partition coefficient, i.e., poor selectivity. Increasing
the selectivity of sorption increases the duration of the elution process
and decreases the concentration of the substance being isolated in the
eluate. In contrast, a frontal process may, indeed must, operate with high
selectivity. Moreover, sorbents have recently been obtained whose
constants of selective ion-exchange sorption for organic substances are in
the thousands. This certainly compensates for the absence of repeated
action and renders the process economically viable because the quantity of
sorbed material may, in some processes that have been developed, reach the
weight of the sorbent. However, frontal chromatography may only be successfully used to isolate one, two, or three components from a complicated
mixture.

Large-scale preparative ion-exchange based on a dynamic equilibrium
analysis and using highly specific sorbents can be used to treat PAS's at a
limited, usually quite slow, throughput of solution. An example of a high-
efficiency frontal separation of two substances close in properties[17] is
the sequential frontal displacement of the amylase and protease produced by
Asp. terricola from the biosorbent Biocarb (Figure 1.3). A gel-chromato-
graph column 100 cm high and filled with the biogel D-50 could only achieve
a partial separation of the same components. It should be noted that the
sorbate-sorbent ratio in a frontal process exceeds the ratio used in
gel chromatography approximately a 100-fold. Obviously modern gel chromatography can achieve a complete separation but only by using a small
sorbate-sorbent ratio.

The new techniques for increasing the efficiency of preparative
chromatography, speeding up the process, increasing the concentration of
sorbate, and, what is most important, for obtaining finer separations of
closely related substances were opened up by kinetic-dynamic analyses[14,
16]. Two very important successes should be mentioned. The first was the

7

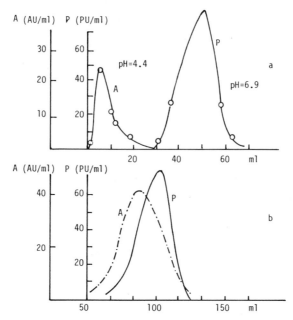

Fig. 1.3. a) Frontal separation on the biosorbent Biocarb-T and b) gel
chromatography on the biogel P-150 for amylase (A) and protease
(P). Mol. mass of amylase 30,000, pI = 4.6; mol. mass of
protease 26,800, pI = 4.3; AU) amylolytic units; PU) proteolytic
units.

analytical solution of kinetic-dynamic problems and the introduction of
criteria that define the conditions in which sharp boundaries between the
substance zones are formed and of a full yield of components obtained
during desorption taking into account the kinetic spreading of the boundaries of the zones. Whereas a quasi-equilibrium regime with symmetric
substance zones (achieved by using small micrometer sized carrier or
sorbent particles) is necessary for highly efficient elution chromatography, large-scale preparative chromatography requires as large a bead as
possible that will, for each system, result in a reasonable approximation
of the quasi-equilibrium and a complete yield in the form of highly concentrated eluates. These criteria, which account for the kinetic and
equilibrium peculiarities of the system, the solution flow rate, and the
bead and column geometry, have shown that for organic ions with molecular
weight between 100 and 1000 daltons, for which the diffusion coefficient
of highly specific and technologically similar ionites is close to 10^{-8}
$cm^2 \cdot s^{-1}$, the length of the diffusion path of the ions must be less than
several tens of micrometers in processes close to quasi-equilibrium, given
a rational flow rate through the column. This analysis enabled another
question to be raised and answered: Could ionites be created that could be
used·in columns without the need to increase the pressure to implement a
quasi-equilibrium dynamic process in frontal chromatography? Surface-layer
and bidispersed ionites are two such successful solutions. In the first,
each bead consists of an impermeable nucleus covered with a layer of sorbed
material; the second uses dispersed micro-bead sorbents (ionites) in inert
highly permeable large beads that have favorable hydrodynamic parameters
for the solution flowing through the column. Surface-layer ionites differ
from pellicular ionites[18] in that their sorbing surface is quite thick
relative to the limiting permissible diffusional path for a quasi-equilibrium process. For the preparative chromatography of many antibiotics,
these sorbents must be spherical beads 100-200 μm in diameter with a

sorbing layer 20-40 μm thick. The sorbed volume in these ionites is 20-30% of the possible sorbed volume of ionites with the same dimensions but made entirely of sorbing material, i.e., without the inert core. Hence it is possible in preparative, and especially industrial, chromatography to absorb from solutions, extracts, and culture liquids quite a large quantity of material per unit volume of the column and yet carry out the process with large solution flow rate and quantitative substance yields. The inclusion of dispersed micro-ionite forms in a permeable inert matrix presupposes that the sorbate can rapidly migrate to the immobilized ionites. We shall consider how to do this successfully below.

It is possible to use, for preparative chromatography, towers filled with a mixture of sorbent beads 20-40 μm in diameter – the dimensions given by the kinetic-dynamic theory – and larger beads made from an inert material such as glass. The solution can be made to flow through such a tower without additional pressure, though it is difficult to charge the tower uniformly. Another difficulty is the gradual redistribution of the ionite and inert filler beads, leading to an impairment in the permeability of the tower. Finally, sectioned columns have been proposed in which relatively thin layers of loosely dispersed beads do not hinder the solution flow.

The regimes and sorbents, which have been predicted theoretically and tested experimentally for quasi-equilibrium frontal chromatography, have enabled enhancements to be made in preparative chromatography and very closely related substances to be separated.

2
Ionite Permeability and Porosity

2.1. TYPES OF HIGHLY PERMEABLE NETWORK POLYELECTROLYTES

The permeability of a network electrolyte for counter-ions, besides its ion-exchange capability, is important for determining whether and how well a synthetic polymeric ionite can be used for the treatment of electrolytes using ion-exchange sorption and chromatography. The permeability of the ionite is very important when the ions of organic or physiologically active substances are involved. This is the topic of this chapter, and we shall consider ionites as networks and deal with their inhomogeneities and their abilities to form porous structures.

It is known that the copolymerization of mono- and divinyl monomers leads to heterogeneous network structures with wide variations in matrix density[19]. Even gel ionites have wide variations between compact and open sections[20]. All the work on creating regular spatial network polyelectrolytes has developed in one direction, that is, to mix together polymer chains and then add bridge-forming components. These interact to form isoporous ionites[21-30]. The permeability of an isoporous ionite, which are gels like most other ionites, is determined by the dimensions of the cells formed by the elements of the network. The networks contain both chemical bonds (via crosslinking agents) and physical bonds (entangled linear polymers and weakly, but not covalently, interacting monomer segments). It is typical of gel ionites that their permeabilities are reduced as the quantity of the polyvinyl component in the copolymer is increased. When dehydrated, gel ionites basically form glassy structures which are impermeable even to gases.

Along with the desire to achieve the greatest homogeneity there is also the desire to create very heterogeneous network copolymers (polyelectrolytes included). Maximum heterogeneity occurs in the macroporous ionites[31-37] that are obtained by copolymerizing mono- and polyvinyl monomers with relatively more of the latter and in the presence of a pore-forming diluent or porogen, usually a precipitant. The term "pore" in the context of these network ionites corresponds to the presence of spaces filled with gas or solvent between the densely crosslinked sections. Copolymers of styrene (St) and divinylbenzene (DVB), and neutral Spheron (Separon) are structural analogs of the porous polyelectrolytes and the basic materials for synthesizing them[36,38-42]. We should note, however, that most macroporous ionites are not absolutely rigid materials with

constant pore dimensions. They can swell to some degree, and this corresponds to changes in the radii and volumes of the pores (and voids) when
an ionite is dehydrated. As regards permeability to counter-ions, differences have been observed between the transport of ions in the pore canals
with sorption on the interior surfaces and migration of the counter-ions
across the densely linked sections. The ability of small metal ions to
permeate the compact sections of macroporous ionites is generally accepted,
while macromolecules (e.g., biopolymers) can only permeate and sorb in the
canals and voids given particular ratios between the dimensions of the
ionite pores and the dimensions of the hydrated macromolecules. As regards
smaller sized ions the probability that they can permeate through a "wall
section" depends on the size of the counter-ion, and the properties of the
ionite (the density of the highly crosslinked areas). Usually the ions of
antibiotics, whose molecular masses are 400-600 daltons, are able to permeate through the "wall section," although the process is slow and only
some of the bound sites in the walls of the macroporous structure are
completely occupied by the organic counter-ions.

Heterogeneous nets lie between gels and macroporous structures and are
permeable to certain macromolecules[16]. When making such ionites it is
necessary to get a blend of permeability for macromolecules (especially
biopolymers), ability to sorb complex counter-ions selectively, reversibility of sorption, and high mass-transfer rates. These intermediate
ionite types are sometimes called density heterogeneous or heteroporous
networks. We suggest that they be called heteronet (or heteroreticular)
because when they are dehydrated they lose most of their porosity. Whether
it is a gel, macroporous, or heteronet ionite that is created depends on
both the percentage of crosslinking agent added during the copolymerization
and on the quantity of precipitant (pore former or porogen). Increasing
the quantity of precipitant in the reaction medium takes the product from a
gel to a macroporous ionite, or to a heteronet ionite. The properties of
the latter two are also aided by adding large quantities of the polyvinyl
component.

The appearance of the density heterogeneity of network ionites (including polyelectrolytes) is due to the creation of initial compact nuclei
structures or globules with various dimensions[38,43]. When nuclei,
globules, or clusters up to 10 nm in size form, spaces appear between them
that can be thought of as micropores. With globules between 15 and 50 nm
in size, mesopores are formed, and these voids and canals significantly
contribute to the surface area S and the total volume of voids and pores
W_0. In addition, macropores can be formed by the aggregation of globules
exceeding 50-100 nm and, in some ionites these can become several micrometers in size. We should note that dehydration leads to the disappearance
of the micropores, a reduction in the number of mesopores, and to little
change in the macropores. To study the porosity and permeability of these
network structures, therefore, specimens should be prepared and treated in
different ways. Macropore formation is facilitated by non-solvating
porogens, while micropores basically occur in the presence of solvating
medium[31,32,37,40]. Precipitation copolymerization with the formation of
meso- and macropores sees the formation of microglobules and a secondary
stage in which the globules increase in size without increasing in
number[44].

The basic class of very permeable ionites is obtained by copolymerization using long-chain crosslinking agents with significant distances
between the vinyl groupings; polymethylenebismethacrylamide is used most,
though there are others[38,45-53,55,56]. The polyelectrolytes may be
obtained both as gels and as heteronet or macroporous ionites, depending on
the quantity of crosslinking agent or solvent used in the copolymerization
or modification. A feature of this group of electrolytes is their very

11

high permeability to different organic ions both in their gel and heteronet states. The term "macronet" should be used for networks of polyelectrolytes which contain long-chain crosslinking agents and not for macroporous ionites.

In order to study the permeability of network polyelectrolytes with respect to different counter-ions or neutral molecules, analysis of the maximum binding to fixed ionogenic groups or the saturation of voids and canals by counter-ions and molecules with different molecular masses, and kinetic methods of determining the migration rates of ions and molecules in sorbent grains are used. The terms equilibrium and kinetic ionite permeability are useful, but it should be borne in mind that both are variable quantities. The accessibility of fixed ionogenic groups and the diffusion of counter-ions depend on many factors, particularly the ionic strength of the solution. It is known that the equilibrium and kinetic ionite permeability are reduced as the degree of ionite swelling is reduced when the ionic strength of the solution is increased. An additional difficulty in the analysis of equilibrium ionite permeability occurs when several bound functional groups are covered by one large organic ion[57]. The covering of canals and network cells by counter-ions is observed in a kinetic analysis too and leads to the formation of relatively homogeneous sorptive layers on the surfaces of the ionite particles[58]. Finally, the possible formation of metastable states is quite important in the analysis of the permeability of network polyelectrolytes. However, a suitable choice of conditions does not hinder either a thermodynamic or kinetic analysis of ion-exchange processes[11].

2.2. METHODS FOR STUDYING THE PERMEABILITY AND POROSITY OF NETWORK POLYELECTROLYTES

The existence of different types of network polyelectrolytes which may be permeable in both the less dense and more dense lattice regions, and in the voids, which may contain a vacuum, a solvent, or a gas, requires different methods for analyzing their permeabilities and porosities. In order to get the best picture, a series of methods ought to be used simultaneously.

To analyze the structure of network polymers a series of methods are used that were originally developed for studying the porosity of solids and dispersions. The special methods for studying the structure of polymers and polyelectrolytes are also used. The choice of methods is governed by the problem and individual features of the ionites. Porosity is usually thought of as characteristic of the voids between the network formation. The porosity of dry network polymers is easily measured because the pore walls are clearly defined. The situation is more complex in swollen copolymers, and so they are more often described from the point of view of the network permeability with respect to counter-ions with different molecular masses (sizes). Porosity is not the same as permeability for these network systems (they are only the same for porous solids) because the porosity of gels and heteronet copolymers disappears or is poorly retained on dehydration.

Gel networks are only permeable in the swollen state (in water or some organic solvents). The permeability of network polyelectrolytes can to a certain extent be compared with their coefficients of swelling. This approach is traditional for ordinary gel ionites, the sizes of whose pores are very small (on the order of 1 nm) in the dehydrated state. The sorption of organic ions onto gel ionites depends on their swelling coefficient. Greater sorptive capacity goes with larger swelling coefficients because the mobility of the network chains and the distance between them increases. Significant sorptive capacity occurs if the pore size considerably exceeds the sizes of the sorbate ions in the hydrated state.

Macroporous systems retain a large volume of pores in the dehydrated state. Heteronet polyelectrolytes have sections of low density whose dimensions are comparable with voids in macroporous systems. These low-density spaces are accessible to biopolymers, viruses[61], and other large objects. Macroporous and heteronet polyelectrolytes are permeable by the cells and uniform elements of blood[62].

Polymer ionites are usually characterized by their specific polymer volume (V_o) and pore volume (W_o). The latter is a function of the heterogeneity of the network. Porosity for both macroporous and heteronet ionites is estimated via the pore surface area (S) and the radius of a pore (r).

The characteristic dimensions of heterogeneity of the polymeric networks, observable using the most widespread methods, lie for most heteronet and macroporous ionites between 1 and 10^5 nm. Porous materials with pore radii between 1 and 100 nm are studied using the sorptive (static) method of gas adsorption. X-ray structural analyses can be used for inhomogeneities between 0.1 and 100 nm, electron microscopy is used for pore and dense formations 10-1000 nm in size, while mercury porosimetry can be used to study a wide range of pores up to 10^4 nm in size (for rigid structures).

Specimen preparation deserves attention because many methods (electron microscopy, mercury porosimetry, BET) can only be used on dry specimens. Clearly, when applied to structures that can swell, these methods cannot give a real picture of their permeability in the presence of solvents, and data from indirect methods that characterize the working (swollen) state of the network must be added. In order to retain the structural detail of a network polymer the choice of a rational dehydration regime is very important. It is shown in[63] that when a carboxylic cationite based on St-DVB in the hydrogen form was desiccated its three-dimensional network shrank, and when re-wet its initial swelling value was not regenerated. This effect is strongest in gel and heteronet systems, while it is weakest in macroporous copolymers. Comparing methods of desiccating the macroporous carboxylic cationite KM-2p[64] indicates that lyophilic desiccation changes the specific volume of a specimen the least. The thermodynamic qualities of a solvent influences its interaction with a polymer. In a good solvent the polymer is elastic, while in a poor one it is rigid. This affects the quantity of deformation of the polymer when the solvent is vaporized and an elastic network shrinks more than does a rigid one[40,66].

Small-angle X-ray scattering has shown[67] that interactions with solutions affect the stability of a porous structure even for rigid chain networks, such as St-DVB copolymer. The porosity of this sort of copolymer depends on the initial swelling and may be severely changed, depending on the thermodynamic properties of the solvent in which the copolymer is swollen until it is desiccated. The presence of toluene in the system when it is heated at high temperatures reduces the porosity of the copolymer almost to zero; further heating in the absence of solvent has practically no effect on the copolymer's porosity. The changes in porosity can be interpreted as disappearances of the smaller pores, the structure shrinking due to surface tension engendered by the phase changes when the solvent evaporates, since fragments of the polymer network are quite mobile[68,69]. There are a number of methods for avoiding or reducing the effects of artifacts associated with desiccation[70]. The method of critical points is based on the fact that every substance has a critical temperature above which the liquid phase cannot exist. When a solvent is evaporated as the temperature is raised above the critical point, it is assumed that all the parameters of the spatial structure remain the same. The method of lyophilic drying consists of the freezing of a swollen specimen and then subliming away the solvent it contains in a vacuum; this reduces the strength of the capillary action that occurs in ordinary drying.

Another method of dehydrating a polymer is the inclusion of a solvent that is closely related to the polymer[42] so that the water is replaced by a succession of organic solvents. The final solvent must have low surface tension, high volatility, and must be inert with respect to the polymer (e.g., water – ethanol – ethyl ether). The main demand on all these methods is that the specimen's volume is the same as the polymer's hydrated form and that its microstructural detail is not changed. A comparison[72] of lyophilic drying and solvent exchange when applied to heteronet carboxylic cationites showed that the former produced less distortion in the dimension of net heterogeneity at the supermolecular level (nodules, domains). In order to preserve the structural features of a polymer net that is obtained from a metastable state with a phase separation and stabilized by a solvent interaction, the most important thing is the rate of desiccation. A comparison of the lyophilic drying of the heteronet cationite Biocarb-T and a very slow desiccation (5 months at room temperature) showed that the rate of desiccation affected the morphology of the dehydrated copolymer, which was best retained by lyophilization. In the second case the initially brittle porous body was gradually compacted to a dense glass (Figure 2.1) due to capillary shrinkage. On the basis of these data the most advisable method for preparing samples, especially for electron microscopy, is therefore lyophilic drying[73].

The porous structures of very permeable network polyelectrolytes were successfully studied using electron microscopy. The preparation of amorphous polymer specimens for electron microscopy and the correspondence between the image and the real structure have been the topics of several papers[40,70,74]. Transmission electron microscopy (TEM) is used to study thin sections, less than 120 nm. When preparing ultra-thin sections (30-40 nm) the dehydrated polymer specimen is encased in an epoxy resin such as Araldite to make it rigid. In order to enhance contrast heavy elements, such as iodine, osmium, or uranium, are added as the electron beam is scattered better by them. However, an additional complication arises out of this because the electron beam can sublime iodine molecules. The introduction of heavy metal salts (e.g., uranium acetate) may affect the structure of the polymer network because the polyvalent metal ions can form additional crosslinks[75]. Thus the problems of obtaining good resolution and contrast for transmission microscopy without distorting the structure have not yet been fully solved.

Scanning electron microscopy (SEM) is used for studying the relief of a specimen's surface. A surface is scanned by a beam of electrons, and the secondary electrons that fly back are registered. A lower working potential is used in SEM than is used in TEM. Greater accelerating potentials can cause the microstructure to "melt" due to the larger surface heating. Thanks to a large depth of focus SEM gives an almost three-dimensional image. Contrast from an amorphous porous structure using SEM is obtained by a double spray coating of carbon and gold onto the specimen surface. The finely dispersed carbon powder fills in the surface relief with a thin amorphous layer, while it is preshadowed on top by the layer of gold, which diffuses freely and carries away the charge formed by the powerful electron beam.

Methods for quantitatively analyzing electron micrographs are still inadequately developed. For ordered structures in which the network elements are clearly defined (e.g., the microglobules of macroporous sorbents), the characteristic dimensions of both the network elements and structure, and the voids between them can be estimated statistically[76,

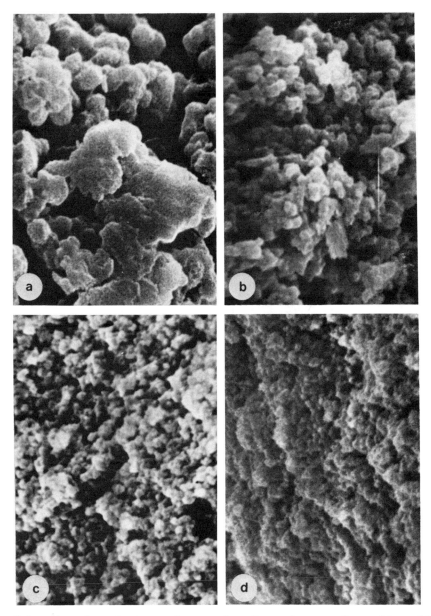

Fig. 2.1. Scanning electron micrographs of copolymers of acrylic acid and
divinylbenzene (SGK-7) for a) 10%, b) 15%, c) 20%, and
d) 30% crosslinking agent (DVB); ⊢————⊣ 1 μm.

p.85,77]. A histogram can be compiled from a large (must be large when
there is a significant spread of sizes) number of microparticle measure-
ments, and this will describe the distribution of microglobules by size.
An average surface area of the network polymer can be calculated from

$$S = \frac{K}{\rho} \frac{\Sigma m_i \cdot d_i^2}{\Sigma m_i \cdot d_i^3} , \qquad (2.1)$$

where m_i is the number of microparticles with diameter d_i,
ρ is the true density, and
K is a form factor, the ratio of surface to volume.

For spheres $K = 6.0$. Particle size distributions and deviations of particle shape from spheres will contribute to errors in the determination of S by this method.

The most informative method for studying the structures of dispersed and porous solids is that of the physical adsorption of vapors and gases. Measuring the specific surface from adsorption data reduced to the calculation of a_m, which is the quantity of adsorbate needed to form a dense monolayer on all the sorbent's surfaces. The BET equation[74;76,p.48] is used to determine a_m for polymolecular adsorption. After several layers of adsorbate have been formed the pores are filled by the capillary condensation mechanism (for pores whose equivalent radii are larger than 2 nm).

In the BET method the quantity of adsorbed gas (usually nitrogen) is measured for dehydrated and degassed samples under deep vacuum and at a temperature close to the gas's boiling point. The BET method only detects significant surfaces in macroporous network copolymers. The surface areas of sorbents with a predominance of small pores (gel polyelectrolytes are one such sorbent because their porosities are reduced by dehydration) cannot be measured by this method. This is because in pores with molecular dimensions the adsorbate condenses in a form close "in volume" to a liquid[76,p.408] and the surface cannot thus be said to be covered by a single layer. The adsorption in micropores is described by the volume filling theory of micropores[78]. The BET method does not determine the surfaces of the micro- or mesopores in heteronet sorbents because they disappear on dehydration. Thus the heteronet cationite Biocarb-T has, according to BET, a surface area of 0.3 m^2/g, while the standard sorbent IRC-50 has one of 1.2-1.8 m^2/g[79]. This is very small compared to the surface area of macroporous sorbents, which can be 60-90 m^2/g [80], and yet heteronet sorbents like Biocarb are more permeable to macromolecules than are macroporous ionites.

A correction to the BET value of S has been introduced[81] for structures consisting of microglobules to correct for the contact area between the globules which the adsorbate gas cannot reach, i.e.,

$$\frac{S^{BET}}{S} = 1 - \frac{n\sigma}{4R} , \qquad (2.2)$$

where R is the radius of a microglobule,
n is the globule's coordination number, and
σ is the adsorbate molecule's diameter (0.43 nm for nitrogen).

The structural sorption method[82] consists in the determination of the quantity of low-molecular-weight liquid vapor that is sorbed by a porous body at different vapor pressures. A sorption-desorption isotherm can then be plotted and the surface distribution and pore volume calculated from it. The sorption isotherms of solvents onto porous sorbents are typically S-shaped with a hystersis loop on the reverse cycle. The method is based on the cylindrical model of a pore. The effective radius of a pore is calculated from Kelvin's equation assuming that the desorption is from a capillary:

$$r_{equ} = - \frac{2V_c \sigma}{RT \ln(P/P_s)} \cos \theta, \qquad (2.3)$$

16

where V_c is the volume of liquid in the capillary,
    σ is the surface tension of the sorbed solvent,
    Θ is the contact angle between a capillary wall and the liquid,
    P is the solvent's vapor pressure, and
    P_s is the solvent's saturation pressure.

For cylindrical capillaries $\Theta = 0$, i.e., the meniscus of the liquid is hemispherical. The quantity r_{equ} characterizes the radii of the concave menisci of the liquid condensed between adsorbed layers on the pore walls.

The application of the sorptive method to porous sorbents encounters several obstacles. Firstly, reliable values for the porosity parameters can only be obtained when the polymer is virtually unswelled by the sorbate vapor, i.e., the degree of swelling is less than 0.5%. On the other hand, when solvent vapors that poorly wet the pore walls are used, reduced values for sorptive volume are obtained. Secondly, at low temperatures the material shrinks due to the temperature, and this is accompanied by a change in the pore structure until the pores completely disappear. The method underestimates the porosity of copolymers, the copolymer of St and DVB[82] for one. It has also been shown[83] that during calculations using the desorption arm of the isotherm, neglecting the influence of the adsorption on capillary evaporation reduces the value for pore radius. Accounting for this effect shifts the differential curves for the volume of mesopores by size in the direction of larger radii. Additional allowances for the change in surface tension with the curvature of the menisci shifts the curves further in the same direction. These shifts are very significant and lead to considerable changes in the porosity parameters and should not be neglected even in rough approximations. Hence without allowing for these factors, which are in fact usually neglected, the sorptive method underestimates the pore values.

The method of solvent sorption, in which the calculation is for the maximum quantity sorbed, does not reflect the true pore volume, and the maximum accessible volume, which depends on the size of the sorbate molecule, is determined.

The pycnometric method is usually used to measure the maximum pore volume W_0^{max} of a polymer network. The quantity W_0^{max} is calculated from the formula

$$W_0^{max} = \frac{1}{d_a} - \frac{1}{d_r}, \tag{2.4}$$

where d_a and d_r are the apparent and real densities of the network. The apparent density is the ratio of the specimen's mass to its volume including pores. For irregular shapes d_a is determined pycnometrically using a liquid that neither wets the polymer's surface nor enters its pores. Mercury at atmospheric pressure is usually used[84]. The true density of a material is the ratio of its mass to its volume excluding the pore volume. In the pycnometric method the true volume is ascertained using a liquid that penetrates the material's pores and wets it, but does not swell the polymer. Pycnometrically obtained values of d_a and d_r may also depend on the nature of the liquids used. In the case of polymer networks, which are known to be formed heterogeneously, the difficulty of this method lies in the determination d_r, the density of the glassy amorphous polymer.

From the definition of true density it must correspond to the value obtained from an X-ray structural analysis, i.e., the X-ray density, which is the ratio of the mass to the volume of a crystal cell. It is very

17

difficult to obtain an X-ray density (obtained at large angles) for the extremely disordered heteronet and macroporous systems because of the intense zones and blurred reflection of the scattered X-rays. Hence the density of a copolymer with the same degree of crosslinking but obtained without the solvent (porogen) is used as the true density[67].

The method of mercury porosimetry is conventionally used to study the porous structure of solids, and assumes that the volume of mercury forced into the material's pores at high pressure is the same as the pore volume. It is known that the wetting contact angle between mercury and the surface of a material with which it does not form an amalgam exceeds 90° (i.e., θ is between 110° and 150°). Hence the permeation of mercury into the pores of such a material (the cylinder model is used here too) encounters a resistive force proportional to the pore radius: $2\pi r\sigma \cos \theta$ (a). In equilibrium, the force with which a column of mercury acts on a pore section can be expressed as $\pi r^2 h_0 \rho g$ (b), where h_0 is the height of the mercury column above the pore. Equating (a) and (b) leads to an equation for the capillary depression which is the same as that for capillary rise, only with a minus sign. The relationship between a pore's equivalent radius and the hydrostatic pressure that must be used to fill the capillary with mercury is given by the equation

$$r_{equ} = - \frac{2\sigma \cos \theta}{P} , \qquad (2.5)$$

where σ is mercury's surface tension,
θ is the wetting angle between mercury and the capillary wall, and
P is the hydrostatic pressure.

Pore sizes are calculated from Equation (2.5), and the volume of open (accessible) pores is calculated from the volume of mercury entering the specimen at different pressures. By graphically differentiating the structural curve $V = f(\log r)$ we get the differential curve $\Delta V/\Delta(\log r)$.

It should be emphasized that mercury porosimetry is based on the cylinder model of a capillary, and so the results are only valid when this model describes the pore structure. Differences between pore volumes obtained by mercury porosimetry and those obtained by the pycnometric method are related to the presence of closed pores or of micropores inaccessible to mercury at a given pressure.

The difficulty with mercury porosimetry is associated with the choice of the numerical values of σ and θ, which not only depend on the nature and cleanliness of the surface, but may also change with pressure. The method is mostly used for comparison. Hence only relative results obtained for a series of copolymers that only differ in one parameter (e.g., degree of crosslinking) will be reasonably reliable. More importantly the method is restricted in application to structures that do not change under external pressure. Comparing mercury porosimetry with inert-solvent sorption indicates that the former overestimates pore dimensions for most polymers because of the reversible elastic deformations of the walls of porous polymers. This effect can be quite considerable in heteronet systems, in which the polymeric skeleton is not very rigid.

Mercury porosimetry is used to study the porosity of dry samples, while small-angle X-ray scattering (SAXS) can be used on polymer networks that may be either dry or swollen by solvent[67,86]. By changing the solvents it is even possible to exclude the influence of hydrophobic interactions on the supermolecular packing of the polymer chains and to determine how changes in the synthesis conditions affect the heterogeneous structures of a network polymer. In small-angle X-ray scattering, beams scattered at small angles of up to a few degrees are measured. If an inhomogeneity on the order of 1-100 nm is encountered, then it appears on the integral curve of the scattering intensity $I(\phi)$. Analyzing these intensities with Guinier's method[87] yields upper and lower estimates of the inhomogeneity. The curve of X-rays scattered by spherical particles uniformly filling a space is close in form to a Gaussian curve, viz.,

$$I(\phi) = I_0 \exp(-kR^2\phi^2), \qquad (2.6)$$

where $I(\phi)$ is the scattering intensity at angle ϕ,
   I_0 is the scattering intensity at $\phi = 0$,
   R is the radius of gyration of the scattering particle, and
   k is a constant including the wavelength of the X-radiation λ,
i.e.,

$$k = \frac{16\pi^2}{3\lambda^2}.$$

A graph of the logarithm of the radiation intensity versus the square of the scattering angle is a straight line with slope $-KR^2$. The radius of gyration of the scattering particles may be calculated from the gradient. Since the inhomogeneity of a heterogeneous network may have different dimensions, the $\log(I)$ versus ϕ^2 plot is nonlinear. The minimum radius of gyration of the scattering particles R_{min} can be estimated from the angle of inclination from the right-hand part of this graph, while the average radius of gyration can be estimated from the angle of inclination in the initial range of the curve (i.e., as $\phi \to 0$).

The linear dimensions of a heterogeneous structure are not unique parameters that can be obtained from SAXS. Porod[87] established a relationship, for an ideal two-phase system, between the integral scattering intensity and the surface area of the scattering sample. A heterogeneous system can be considered ideal if the electron densities in both its phases are constant and if there is an abrupt jump in the electron density at the phase boundary. The intensity of scattering from a macroscopic isotropic sample may then be given as

$$I(S) = VP(1 - P)(\sigma_1 - \sigma_2)^2 \int \gamma(r) \exp(2\pi r_i S) dV, \qquad (2.7)$$

where $\dfrac{2 \sin \phi}{\lambda}$ is the size of the diffractional vector S,

   V is the sample's volume,
   P is the volumetric concentration of the scattering phase in the sample,
   σ_1, σ_2 are the electron densities of the first and second phases, and
   $\gamma(r)$ is Porod's characteristic function.

When measuring the scattering from a heterogeneous network polymer in the absence of solvent, the density of the substance in the pores can be neglected in comparison with that of the polymer matrix. The surface area of the phase interface is given by

$$S/V = 8\pi P(1 - P) \frac{\lim\limits_{S\to\infty} S^3 I(S)}{Q}, \qquad (2.8)$$

where Q is obtained from experimental data by graphically integrating the function

$$Q = \int_0^\infty SI(S) dS. \qquad (2.9)$$

19

For ideal two-phase systems the tail of the log I(S) vs −log S curve must have a slope of −3, and for slot collimation one of −2 (Porod's law).

In the case of macroporous systems the porosity given by SAXS may be less than that given by the pycnometric method because SAXS is not sensitive to large pores (>100 nm) and hence they do not contribute to the scattering intensity[67]. Thus this method is more effective for heteronet systems such as the copolymers of methacrylic acid with different crosslink ratios or synthesized in solvents with different affinities for the copolymer[86]. In the case of gel networks, SAXS, like the BET method, does not detect any interface surfaces[86].

Gel-permeation chromatography, which is based on a molecular sieve effect, has been used several times to study the porous structure of sorbents[90,91]. The advantage of this method is that, like SAXS, it can be used to study the sorbent in its solvated state without its morphology having been changed, this being especially true for heteroporous materials.

Porosimetry based on gel chromatography can be used to find the distribution of pores with respect to radii[92,93]. This method is to determine the partition coefficient (K_p) of a macromolecule like dextran between the sorbent and solution. A theoretical relationship between the partition coefficient and the ratio of the dimension of the linear macromolecule (R) to the pore radius of the sorbent (r), $K_p = f(\frac{R}{r})$, has been established for a dilute solution when the polymer-solvent system is at the Flory temperature and there is no adsorption of the macromolecule onto the sorbent matrix[94]. The partition coefficient of the macromolecules between the phases is then entirely determined by the change in the entropy of conformation that takes place when the molecule moves from the solvent into the sorbent pore. For sorbents with wide pore-radius distributions, the pores of each size contribute to the partition coefficient in proportion to their fraction of the total pore volume. Thus

$$\langle K_p/R \rangle = \int_{r_1}^{r_2} K_p(R/r)\psi(r)dr, \qquad (2.10)$$

where $\psi(r)$ is the pore radius distribution function, and
r_1 and r_2 are the limiting values of the pore radius.

The task reduces to the experimental determination, using gel chromatography, of the partition coefficient of the macromolecule between the phases and the numerical solution of (2.10) for the system.

Until now a description of the real structure of a porous body has been a very complicated task. The cylindrical pore model has been the basis of many research methods. This is an approximation which has been contradicted in many cases by data from electron microscopy, for example. Another well-used model is that of packed spheres, in which the medium is approximated to a set of monodispersed spheres lying in a defined order. The calculation of the porous structure is complicated if the coordination number (the number of contacts a sphere has) increases.

Real systems are significantly different from the ideal models, firstly, because of the polydispersity of the microglobules - the size heterogeneity of the polymer phase's particles - and, secondly, because of the particles differing geometries and irregular spatial distribution; the globules may be distinct or clustered together linked covalently due to chain inter-penetration, which occurs during copolymerization. This can be seen clearly in the electron micrographs of heteroporous polyelectrolytes[64,95].

In order to bring the sphere model closer to reality another parameter (a) has been suggested[96]. This is the average distance between the spherical particles and is chosen so that the values of S and V coincide with the values of these quantities characteristic for the system. This parameter can be used to describe both separate particles (a > 0) and clusters (a < 0).

A theoretical analysis[96] has shown that the packed spheres model can be applied to network polymers if the porosity is greater than 26% (this corresponds to face-centered cubic packing of the spheres with n = 12). The greatest porosity (66%) corresponds to n = 14 (tetrahedral packing). The cylindrical pore approximation is valid for systems whose porosities are greater than 60%. The value of porosities between 30 and 78% is inversely proportional to the coordination number of the packing. For n = 2 the structure is not stable.

Since each structural method of investigation has its own limitations, a detailed description of the heterogeneity of a porous sorbent must come from a whole series of structural analysis methods and from a comparison with data from indirect methods based on volumetric processes.

A study of the kinetics of ion exchange with a pair of ions can be used to estimate the hindrance to their migration within ionite beads. It should be noted that the kinetic method (like every other method) has some limitations in its application to the study of the permeability of network ionites. However, it can yield additional information about ionite porosity and permeability. Only those kinetic processes for which the limiting process is diffusion (quasi-diffusion) into the ionite beads can be used to solve this problem. However, difficulties then arise when analyzing network permeabilities due to the mechanism of ion migration. Diffusion can be imagined as the flow in a capillary of a concentrated solution fluid situated in a cell that creates extra pressure due to the elastic forces in a polmyer system that can be deformed by dehydration and the ionic environment. It is also possible to consider diffusion as the migration of ions in a drop of concentrated solution[11,97,98], or as the substitution of a series of functional groups one after the other by a counter-ion. Both processes must be the result of the movement of both co-ions and counter-ions[99,100]. Other opinions on the mechanism of ion transfer and exchange will be considered later. This all indicates that the kinetic permeability of ionites is a complex phenomenon in which the structural density and cell dimensions of a network are but one, albeit important, aspect of the phenomena dealt with under ion migration.

In order to study the kinetic permeability of an ionite not only must the process's limiting stage be determined, but the conditions must also be listed in which the ion-exchange kinetics are defined by gel diffusion (ion diffusion within the sorbent beads) and not by film diffusion (diffusion to or from the beads). For most systems studied (if the complex-forming ionites are not considered) chemical limitation, when the slowest step is the interaction of an counter-ion and a functional group of the ionite, is rarely observed. We can, however, assume that the importance of this step is still poorly understood for several organic counter-ions.

In order to appraise gel limitations, the absence of any influence exerted by the ion concentration in the solution and by migration on the ion-exchange kinetics[101-104] should be considered first. Experimental kinetic curves[11,97,104,105] are also used. In the case of gel limitation a linear relationship between the extent of exchange and the square root of time is observed. Further proof of this step being limiting comes from an analysis of the Adamson-Grossman-Helfferich criteria[105,106] and the Biot criterion[104]. Finally, a widely accepted method of elucidating limiting steps is to break the continuity of the phase contact. If the phase contact is re-established and a kink appears in the kinetic curve after the break, then ion diffusion in the ionite beads is the limiting stage in the establishment of equilibrium. It should be emphasized that to do a rigorous analysis of the kinetics of ion exchange requires a comparison of the results of different limiting-step methods. If only one experiment with a broken phase contact is considered, erroneous conclusions about the relative contributions of external (film) and internal (gel) diffusions may be reached.

To calculate the diffusion coefficients the asymptotic expression for the initial section of the kinetic curve in the case of spherical particles[107] that can be filled by an counter-ion is usually used, viz.,

$$F = \frac{6}{R} \sqrt{Dt/\pi},\qquad(2.11)$$

where F is the extent of exchange,
t is time,
R is the radius of a bead, and
D is the coefficient of internal diffusion.

Bearing in mind that ion exchange can be delayed by the adsorption of large ions, it is best to evaluate an average coefficient of diffusion[108], i.e.,

$$\bar{t} = R^2/15D_m,\qquad(2.12)$$

where the average time $\bar{t}$ is evaluated as the first primary statistical moment of the kinetic curve

$$\bar{t} = \int_0^1 t\,dF.\qquad(2.13)$$

The observed fall off in the rate of ion diffusion within the ionite beads[109-111] as the ionite is filled with organic counter-ions can be interpreted in different ways. One possibility is that it is due to the difficulty of moving within a network structure filled with large ions that reduce the size of the cells the ions travel through. The second often frequently encountered feature is the heterogeneity of network polyelectrolyte beads. Finally, the third feature of the diffusion that could reduce counter-ion migration is the limited width of the canals or voids along which the ions move from the periphery of a bead to its center.

Hindrances to the diffusion of organic counter-ions in network systems may cause the sharp diffusional front that is observed during the sorption of proteins by the very permeable heteronet ionites[112]. The heterogeneous final distribution of ionites with respect to ionite radii[112-114] may be due to the same reasons. An extrapolation of the rate of counter-ion diffusion in network structures is then possible[114]. The limit of this process, which leads to a homogeneous external layer of sorbent and an internal area free from organic counter-ions (the cloud-nucleus model), can be considered assuming that the counter-ion diffusion coefficient is constant in the outer accessible cloud[112-114]. The coefficient can then be calculated for the linear section of $F-\sqrt{t}$ from the equation

$$F = \frac{6}{1 + \rho + \rho^2} \sqrt{Dt/\pi \ell^2},\qquad(2.14)$$

where $\rho = 1 - \ell/R$, and
ℓ is the thickness of the diffusion layer.

A comparison of the equilibrium and kinetic methods of investigating network polyelectrolyte permeability, and a study of their porosity and heterogeneity makes possible not only a description of the structure of rigid macroporous ionites, but also an evaluation of the basic structural elements of heteronet, macronet, and gel ionites.

2.3. THE STRUCTURE AND PROPERTIES OF VERY PERMEABLE NETWORK POLYELECTROLYTES

The most significant application of very permeable ionites in preparative chromatography is in the isolation, separation, and purification of large electrolyte ions, especially those of physiologically active substances[11,15,115,116]. In order for them to be successfully applied the ionites must possess a series of properties that are crucial for elution. High permeability (equilibrium, quasi-equilibrium, and kinectic) must be combined with thermodynamic selectivity and reversible sorption of complex ions. Columns filled with the ionite must have good hydrodynamic properties such as adequate solution flow rates without using high pressures. Very permeable ionites based on Sephadex (and biogels) cannot be applied successfully in this way because the weak additional interaction between the ionite matrices and organic counter-ions leads to poor selectivity, which appears as a small ion-exchange coefficient, i.e., there is a sharp reduction in the sorption capacity with respect to organic ions in the presence of mineral electrolytes[117]. It is well known that the highly permeable gel ionites do not to a great extent sorb large organic ions, especially proteins, reversibly. Sorption irreversibility is associated with the rearrangement of the structure of the ionites as large ions are sorbed and the formation of polymer-protein complexes with large interaction energies. The sorption irreversibility is also promoted by the coarsening of the copolymer network close to the sorbed ion; this naturally hinders the diffusion of the bonded ion during desorption. A successful solution of the problem of irreversible sorption of complex organic ions in permeable ionites (along with very selective ion sorption) is the use of structurally (or osmotically) stable ionites based on heteronet structures[115,116]. The feature of these ionites is that their degree of swelling is little changed by alterations in pH and ionic strength of the solution, although when dehydrated they lose most of their permeability. It should be noted that the high sorption reversibility of organic ions by ion-exchange Sephadexes and biogels is not due to structural stability (it is absent) but to the low interaction energy; additional interactions besides the electrovalent one are absent. It is very important that the structural (osmotic) stability of an ionite is obtained by using a large concentration of crosslinking agent. This provides satisfactory conditions for elution in columns because of the small tendency to deform under the restricted pressure necessary as the solution elutes down the column.

All this stresses the great value of heteronet ionites, including those macronet ionites which can be classified in this group. Further consideration of the best method of analyzing permeable structures and their properties will be for this class of network polyelectrolytes.

Supermolecular structures form within network polyelectrolytes during phase separation and the packing of domains in microglobules[119], and they have been observed in ultra-thin sections of network carboxylic acid based polyelectrolytes, high-resolution microscopy revealing structures on the order of 5-10 nm[66,72]. It is at this level of polymer network organization that capillary contraction forces, which cause polymer volume reduction upon dehydration, operate[63].

Electron micrograph pictures of gel (micronet) sorbents at the fractional micrometer level detect only continuous polymer phases, while both heteronet and macroporous sorbents are in the form of aggregates of microglobules[95,123]. Microglobules are the smallest element of the network structure that can be observed with SEM. The porous structure of macroporous and macronet sorbents is formed by the voids between the aggregates (the meso- and macropores). Pictures obtained at different stages of precipitation copolymerization can be used to trace the formation and packing of microglobules as the final structure of a polymer[125-127].

The electron microscopic analysis of polymer networks at different magnifications shows up the heterogeneity of their structures at the different levels, each of which can be highlighted using a different structural investigation. The size of a macropore can be determined using TEM at small magnifications, while the packing of globule formations in aggregates can be established with SEM. This adds information about the morphology of bead surfaces and their internal domains (edges and surfaces) to the data from TEM which are obtained from thin sections.

Thus the supermolecular packing of polymer networks determines the heterogeneity of network polyelectrolytes obtained from precipitation copolymerization (Figure 2.1). In heteronet sorbents much of the interface and total pore volume is due to the organization of the structure's elements into three supermolecular levels, viz., domains, microglobules, and microglobule aggregates. The formation of a copolymer's structure in the presence of solvents is explained by Dusek on the basis of the classical Flory-Huggins theory of solutions[128,129]. According to Dusek's thermodynamic model[130], the chief factors affecting phase separation during a heterophase copolymerization are the degree of crosslinking of the three-dimensional polymer, the solvent concentration, and the interaction energy in the system, i.e., the Huggins constant χ for the polymer-solvent pair. Dusek's theory yields a critical degree of polymerization at which phase separation occurs, and describes the equilibrium between the phases. The complexity of studying copolymerization patterns in the presence of solvents arises because the monomer itself acts as the diluent and the concentrations of the polymer and monomer are functions of the degree of polymerization.

At present the "homogeneous nucleation" mechanism of phase separation, i.e., the formation of a copolymer "stock" (monomer segments lined to form oligomers) as the first stage of precipitation copolymerization, is widely accepted. When the phase separation is slow and takes place close to the thermodynamic equilibrium between the polymer and precipitant phases, the polymer network is formed from relatively similar supermolecular formations which, even in the expanded gel state, are like the fluctuations within a single phase. Heterophase fluctuations can be substantially increased during phase separation by taking the system away from equilibrium (by increasing the degree of supersaturation)[37,131]. The mechanism and kinetics of the formation of a three-dimensional copolymer phase are in this sense closely linked. Under non-equilibrium conditions very large fluctuations in density arise which are fixed by chemical linkages, and metastable structures are formed which cannot be obtained under equilibrium conditions. The formation of macroporous copolymers relies in the first place on the role of solvents, i.e., the phases separate due to changes in the solvation coefficients of the systems components and net macroporosity ensues[37]. Porous copolymers synthesized in the presence of non-solvating solvents are very dispersed structures. This means that the copolymer cannot dissolve in, and is not swollen by, the mixture of unreacted monomer and porogen as it is polymerized. As a result, the system becomes super-saturated with respect to the coploymer, and the new polymer phase separates from it in the form of dispersed particles. An inert solvent used during precipitation copolymerization is traditionally called a porogen for systems like styrene (St) and divinylbenzene (DVB). For flexible-chain systems based on acrylic acids (AKA) or methacrylic acids (MAA), the term copolymer precipitant is used.

A solvent used to form a porous structure may have a different affinity for the polymer network: it may be solvating or non-solvating[31,32,

40,132]. In the first case - the copolymer's chains are solvated - very small microglobules (on the order of nanometers) are formed with the distances between them of the same order, i.e., micropores are formed, all of this being due to the low surface tension. For the St-DVB system octane is one such solvating solvent; water is one for AKA, MAA, and divinyl copolymers. A small reduction in the quantity of solvating solvent increases the surface tension, i.e., the dispersed polymerized phase is compressed with the formation of microparticles. This effect is clearer with small quantities of crosslinking agent because with a large amount of chemical bridging the compression of the microglobules being separated is reduced.

In the second case - the solvent does not solvate the copolymer's chains and, in particular, does not wet the surface (e.g., the copolymer of MAA with heptane as the precipitator) - the initial surface cannot be retained and the microglobules emerging from the phase separation are compressed and agglomerated. As a result, very large agglomerates remain in the network, and when packed together they leave large pores between them.

The mechanism by which porous structures based on St-DVB are formed is described in detail[35,42,132,137]. It has been shown that crosslinking agents give stability to the porous structure formed with respect to a change or removal of solvent. Copolymers acquire porosity, and retain it after dehydration, providing the quantity of solvent and crosslinking agent exceeds a critical value, which is characteristic for each polymer system[32]. It has been established that given a constant concentration of crosslinking agent, increasing the concentration of porogen in the polymerizing system continuously increases the total pore volume, while the surface area passes through a maximum. This has been established experimentally for the St-DVB system for a range of solvents[1,32,138,139].

Increasing the concentration of crosslinking agent in the initial monomer mixture reduces the solubility (and swelling) of the polymer which separates out, i.e., the supersaturation (divergence of the system's metastable state from thermodynamic equilibrium) grows. A statistical description of the gelation of the chains at the earlier stages of copolymer separation[140] assumes the point size of a chemical bridge. However, when complicated crosslinking agents are used, neither the size nor the flexibility of the latter, e.g., polymethylenebismethacrylamide or hexahydro-1,3,5-triacryloyltriazine (HTA), can be neglected. A molecule of HTA (and its analogs) is 11 Å in size[141], and gelation with such a crosslink results in a very heterogeneous network, the heterogeneity growing with the increase of the concentration of crosslinking agent in the system.

The mechanism of forming porous structures from flexible, chain polymers (AKA and MAA copolymers) linked by di- and trivinyl crosslinking agents depends on both the nature of the copolymer precipitant and the quantity of crosslinking agent[120,121].

As we have shown, chain gelation at the initial stages of the reaction increases the heterogeneity of the system with low rates of conversion[140]. When the crosslinking agent concentration is increased, this gelation yields densely crosslinked compact gel microglobules whose surfaces have residual double bonds, which link together globules in the final stages, to form the network's wall structures. A study of the morphology of the AKA-DVB copolymer given different crosslink ratios has shown that with 10% DVB in the system, a porous structure is formed whose heterogeneity dimension is on the order of 1 μm. At higher concentrations of DVB this dimension is reduced, and with 20 to 30% DVB the dimension is even smaller, on the order of 0.1-0.2 μm. Apparently with the denser crosslinking the microglobules from the initial copolymerization stages have a large number of double bonds at their surfaces and aggregate faster than the individual microglobules grow. At low concentrations of crosslinking agent the growth of each microglobule is slower, and only after they have reached a certain size do they combine and form a common structure. Similar relationships between porosity and crosslink ratio have been observed in glycidyl methacrylate and ethylene glycol dimethacrylate copolymers[142]. A change in the supermolecular packing of the polymer network when the crosslink ratio is increased - from gel to macroporous systems - has been observed over the series of KMDM-type carboxylic acid based cationites[120] in which N,N'-ethylenebismethacrylamide was used as the crosslinking agent.

Structurally stable polymer networks (with respect to osmotic changes in the media, viz., pH and ionic strength of the external solution) are obtained by the precipitation copolymerization of MA with the trivinyl crosslinking agent HTA[16,88,121,143]. The synthesis is conducted in a mixed aqueous organic solution. Reducing the "quality" of the solvent - by going from 30% to 5% acetic acid in the MAA-HTA system - increases the supersaturation, and in the end increases the surface area S of the copolymer[88,121]. The extreme nature of the dependence of S on the quantity of HTA in the copolymer is probably related to the redistribution in size of the pores toward larger values. This seems also to explain the non-trivial dependence of the MAA-HTA copolymer's specific volume V_{sp} on the crosslink ratio, i.e., as the concentration of HTA is increased in the copolymer, the specific volume gets larger, as does the swelling of H-form ionites (Figure 2.2). A similar effect is noted when unsaturated aromatic acids, such as the methacryl derivatives of para-aminosalicylic or benzoic acids, are added. Increasing the HTA concentration (at some levels) raises the copolymer's V_{sp}, and the volume change when the carboxylic groups are ionized is reduced. This may be related to the increased rigidity of the polymer network. All these effects may be explained by phase separation during the precipitation copolymerization when the concentration of crosslinking agent in the monomer mix is increased and by the very non-uniform distribution of it in the copolymer bulk. A consequence of this is that the polymer network obtained has a very large internal surface. Polyfunctional crosslinking agents increase the heterogeneity of the networks because of the formation of covalent bonds between microglobules formed at earlier stages and the formation of globule aggregates. Heteronet systems obtained by heterophase copolymerization using solvating solvents are intermediate between gel and macroporous systems. We shall show below that the metastable polymer structures obtained during the phase separation stage and fixed by chemical bridges retain an excess of free energy whenever solvent and counter-ion are subsequently exchanged. The degree of

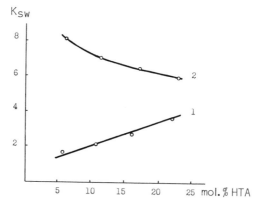

Fig. 2.2.   Volume swelling of Biocarb-S versus crosslinking agent concentration: 1) hydrogen and 2) hydrogen-sodium form ($\alpha \sim 0.5$).

heterogeneity in heteronet polyelectrolytes is shown in Figure 2.1 as a function of the crosslink ratio.

Heteronet anionites[146] obtained from precipitation copolymerization using HTA are similar to heteronet cationites in a number of basic ways. The sorption of proteins (serum albumin or insulin) onto these network polyelectrolytes occurs with a large sorption capacity at all crosslink ratios (Figure 2.3). A low concentration of HTA yields a gel ionite, while 15-24% crosslinking agent results in a clearly heterogeneous network structure. The high adsorptive capacity of heteronet anionites with large crosslink ratios is especially valuable because the sorption is then completely reversible. Another adsorption pattern of macromolecules by heteronet anionites is observed when the ions are not globular but flexible coils. Heparin is only sorbed with great capacity when the anionite has very heterogeneous structure; gel anionites obtained with small crosslink ratios are virtually impenetrable to heparin (Figure 2.4). The degree of swelling of a heteronet anionite correlates with its penetrability for globular protein macromolecules (Figure 2.3). In contrast, the flexible polyelectrolyte coils of heparin, which has the same molecular mass as proteins, cannot penetrate even the most open gel network.

Macronet ionites with N,N'-polymethylenebismethacrylamide as crosslinking agent have been studied from the point of view of kinetic penetrability for various organic ions. A sharp growth in the diffusion coefficients of tetracycline in sulfocationites is observed when the usual gel St-DVB copolymers are substituted for macronet sulfopolyelectrolytes containing ethylene-, hexamethylene-, or decamethylenebismethacrylamides[147] (Figure 2.5). A rise in the quantity of crosslinking agent introduced reduces the permeability of standard ionites but has relatively little effect on the kinetic permeability of macronet ionites with respect

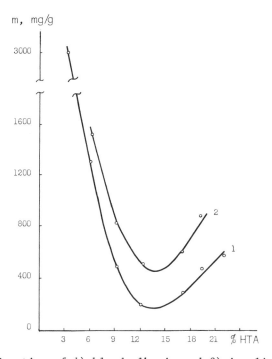

Fig. 2.3. Sorption of 1) blood albumin and 2) insulin on a heteronet anionite.

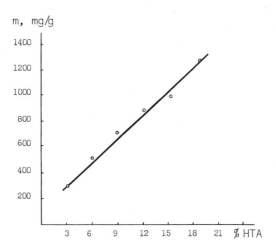

Fig. 2.4. Sorption of heparin on a heteronet anionite.

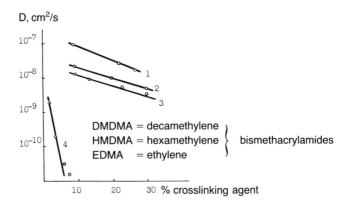

Fig. 2.5. Diffusivities of tetracycline in macronet ionite beads. Crosslinking agent: 1) DMDMA; 2) HMDMA; 3) EDMA; 4) Dowex-50B.

to the ions in question. A similar relationship has been obtained for the
diffusion of triethylbenzylammonium ions in macronet ionites, the diffusion
coefficient being ten times greater than it is for diffusion through
standard gel ionite beads with the same swelling coefficient[148]. The
universal nature of the high kinetic permeability of macronet ionites was
corroborated when the sorption of streptomycin on carboxylic cationites was
studied[111]. Increasing the number of methylene links in the crosslinking
molecule usually leads to a more permeable ionite. However, going from
ethylenebismethacrylamide to hexamethylenebismethacrylamide increased the
kinetic permeability of carboxylic macronet ionites with respect to streptomycin; moreover, the diffusion coefficient fell off if the ionite
included decamethylenebismethacrylamide. It can be assumed that deca-
methylene units in carboxylic cationites create more polymer entanglements,
possibly via hydrophobic interaction, without a reduction in the hydration
and swelling of the network polyelectrolyte. The high kinetic permeability
of macronet ionites at swelling coefficients observed in standard gel
ionites indicates that there is a considerable heterogeneity in the network
structure. At high crosslink ratios macronet ionites become similar to
heteronet polyelectrolytes with respect to permeability.

The exceedingly high kinetic permeability when the quantity of divinyl
component (polymethylenebismethacrylamide) in the macronet copolymer is

large must be considered as the combination of very open sections, through which the counter-ions can diffuse quickly, and small dense regions in which the diffusion coefficient is lower but for which the diffusional path is very short. This all leads to good kinetic parameters during sorption. Thus kinetic permeability cannot be interpreted as a simple function of the network crosslinking in a polyelectrolyte. The idea that the rate and selectivity of counter-ion sorption are related has even been developed in a number of papers[15,150], with an increase in rate leading to a decrease in selectivity in these systems. This assumes a role for the free energy (or enthalpy) of the counter-ion/bound ion interaction in the activation energy of ion migration in the ionite. However, in macronet ionites a parallel growth in both selectivity and rate of sorption of organic counter-ions has been observed[151,152]. This makes another model for describing kinetic permeability more attractive. In this model the high selectivity of organic ion sorption is determined to a considerable degree by the growth of the entropy of the system[11,117]. It can be assumed that organic counter-ions migrate through a macronet ionite without completely disrupting their interactions with the ionite matrix when the combined electrovalent and additional interactions alternate with only a weak interaction between a counter-ion and the polyelectrolyte matrix. This reduces the activation energy of quasi-diffusion and corresponds to a high migration speed with large sorption selectivity. Highly permeable ionites can also be obtained using other crosslinking agents with widely spaced vinyl residues. These include the diacrylic esters of hydroquinone, of Bisphenol A, or of di(p-hydroxyphenyl)diphenylmethane[153,154], and the mono-, di-, and triethylene glycol dimethacrylates[155,156]. The kinetic permeability of macroporous ionites is straightforward for ions that can freely diffuse into the macropores but cannot penetrate into the compact regions. Thus the quasi-diffusion coefficient of insulin in macroporous ionites is two orders of magnitude higher than that in gel ionites.

Investigations of the hydration of network polyelectrolytes have provided information about their permeabilities, porosities, and inter-actions between the network elements and the state of absorbed water[104, 157,158,160-165]. Some heteronet and macronet ionites have minima in their curves of swelling factor versus concentration of crosslinking agent, while those of most other ionites do not[66,166]. If the swelling of a loosely crosslinked ionite of this kind produces a gel structure, then the swelling of a densely crosslinked heteronet depends on the swelling of the more loosely crosslinked regions. This results in the formation of canals that are easily penetrated by large ions, and of voids that are observable using electron microscopy and small-angle X-ray scattering[89,95]. The large degrees of swelling of these ionites is in itself evidence that all the permeable regions are not simply free space filled with solvent, but some are swollen loosely crosslinked regions. This is one of the signifi-cant differences between heteronet and macroporous polyelectrolytes, the latter having only free pore space filled with solvent or gas. Heteronet polyelectrolytes obtained by heterophase copolymerization with large cross-linker concentrations may have another important property that clearly distinguishes them from gel ionites, namely, their swelling is not much affected by changes in the solution's pH or ionic strength, i.e., they are osmotically and structurally stable. However, they lose most of their permeable and quasi-porous structure when dehydrated by desiccation. The isopiestic isotherms showing how water absorption depends on the vapor pressure for ionites with different crosslink ratios intersect each other [161,169]. For gel ionites they intersect at small water vapor pressures, while for macronet ionites they intersect at large water vapor pressures. The differential thermodynamic potential decreases in proportion to the water absorbed, and beyond a particular point the sorbed water is no longer strongly bonded to the ionite's structure (the solvation energy is small). This is usually called "free" water. The relative quantity of free water is quite large for heteronet and macroporous ionites.

2.4. THE CONFORMATIONAL STATE AND FLEXIBILITY OF THE STRUCTURAL ELEMENTS OF NETWORK POLYELECTROLYTES

The structural aspects of studying network polyelectrolytes should be considered with respect to whether the state of the network's structural elements can be changed and to the mobility of their elements. Until recently the conformational state of the polyelectrolyte chains between chemical crosslinks was poorly studied, as was the mobility of sections of the network. For polyelectrolytes the presence of charges in the structure and the consequent electrostatic interactions must be taken into account. Electrostatic interactions between monomer bridges are dominant in strong polyelectrolytes, whereas in weak polyelectrolytes the non-electrostatic monomer interactions must also be considered. This set of interactions also occurs in changes of the structural and dynamic characteristics, i.e., the mobilities of polyelectrolyte chains. Naturally in weak polyelectrolytes this is substantially dependent on their ionization.

From the point of view of the conformation of network polyelectrolytes (NPE's) and changes in conformation, the linear polymethacrylic acids (PMAA's) and the networks based on them have been studied the most. Ionization in linear polyelectrolytes can be studied by viscometry, light scattering, polarized luminescence, and potentiometric titration. The results so far obtained have established that PMAA may exist in one of two conformations, depending on the degree of ionization. In the un-ionized state local (secondary) structures are formed in the PMAA due to hydrophobic interactions between the methyl groups of the methacrylic acids (MAA) and hydrogen bonds between the un-ionized carboxylic groups. The secondary structure is retained at low levels of ionization $\alpha = 0-0.15$, but is disrupted due to the electrostatic repulsions of the ionogenic groups when $\alpha > 0.15$. The disruption of the secondary structure is accompanied by the cooperative transition of the PMAA from an ordered conformation to a disordered one and a change in the polymer chains' mobility. The polarized luminescence method yields exhaustive information about intramolecular mobility for linear polyelectrolytes[170]. This method is based on the use of covalently attached luminescent groups.

A study of the molecular relaxation parameters of network polymer systems is of considerable interest both to establish how the network is formed and to study how a NPE functions as a sorbent of large organic ions. It has been suggested that insoluble NPE's be studied as stable finely dispersed suspensions in water. A finely dispersed suspension of a carboxylic macronet polyelectrolyte – a MAA copolymer with 2.5 mol.% N,N'-ethylenebismethacrylamide (EDMA) as the crosslinking agent – was used to study changes in the mobility of polymer chains between the chemical (or physical) bonds[172]. The copolymer was obtained by bulk radical polymerization, and it was ground, fractionated, and carefully washed free of lower-molecular-weight components. The stable nonsettling fraction, which had particle dimensions of 1 μm, was used; we shall call it PMAA-S. The particle dimensions were estimated by turbidimetry. In order to study the suspension by polarized luminescence, luminescing anthrylacyloxymethane groups were attached to the carboxylic acid groups of the network (one to every 1000 or so monomer units)[171].

The ionization of a PMAA-S suspension was studied by polarized luminescence[172]. The changes in the relaxation time, which characterizes the
intramolecular mobility of polymer chain fragments, during the ionization
of a network and linear polyelectrolyte (based on PMAA and PAA) are compared in Figure 2.6. The monotonic change in the network relaxation time
of a polyelectrolyte containing methacrylic acid differs from the sharp
rise in the relaxation time in the case of PMAA. The cooperative conformational transition is observed for the PMAA and PMAA-S at the same
values of $\bar{\alpha}$. Experimental studies of these systems using potentiometric
titration, which are presented below, confirmed the presence of a conformational change in the PMAA-S fragments. A sensitive indicator of the
presence of structured sections in a linear PMAA (at $\bar{\alpha} < 0.12$) is the
ability of the structured form to bond acridine orange dye when its concentration in the solution is low (1 dye molecule per 300, or so, monomers)[173]. The reciprocal of the polarized luminescence of acridine
orange in water with $\bar{\alpha} = 0$ are the same for PMAA and PMAA-S[172]. The
ability of PMAA-S in its un-ionized form to interact with acridine orange
is reduced as the carboxylic groups are ionized. This means that during
the transition from the un-ionized to an ionized state there is a disruption of the local structure that is present in the un-ionized form of
PMAA-S.

It should be noted that the change in the intramolecular mobility of
the NPE's chains ($\bar{\alpha} \sim 0.1$-0.4) is accompanied by a significant change in
the swelling factor. It is clear from Figure 2.7 that the volumetric
swelling grows rapidly and reaches the limiting value set by the network
skeleton at $\bar{\alpha} \sim 0.4$. The use of a finely dispersed fractionated suspension
of PMAA-S enables carboxylic NPE to be directly titrated potentio-
metrically[174]. The titration of large pieces (~ 100 µm) of carboxylic
acid NPE's is characterized by the slow establishment of the equilibrium of
Na^+-H^+ exchange (several days)[175]. In contrast, the small size of the
particles (~ 1 µm) eliminates these kinetic difficulties when the PMAA-S

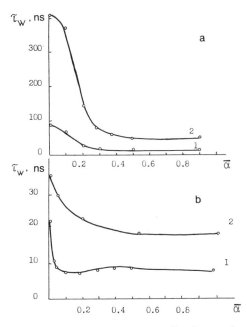

Fig. 2.6. Relaxation time τ_w versus polyelectrolyte ionization.
a: 1) Linear PMAA; 2) network PMAA-S. b: 1) Linear PAA;
2) network PAA-S.

31

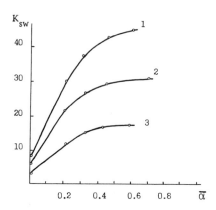

Fig. 2.7. Swelling of macronet carboxylic cationites during ionization.
Copolymers of MAA and EDMA in the presence of 30% acetic acid
in slab form. 1) 1 mol.% EDMA; 2) 1.5 mol.% EDMA; 3) 2.5
mol.% EDMA.

suspension is neutralized with alkali, and enables the ionization process
to be studied by potentiometric titration.

The two possible conformations of linear PMAA dependent on ionization
show up clearly in the potentiometric titration curves[176]. The curves
for a network (curve 1) and a linear (curve 2) sample are given in Figure
2.8 with $pK_{\overline{\alpha}}$ versus $\overline{\alpha}$ (where $pK_{\overline{\alpha}} = pH + \dfrac{1 - \overline{\alpha}}{\overline{\alpha}}$). The shapes of the two
curves are the same. By comparing them with data about intramolecular
mobility (Figure 2.6) it is possible to estimate the state of the polymer
chains at different sections of the curves. For $\overline{\alpha}$ = 0-0.15 the structured
form of PMAA-S is being titrated, while $\overline{\alpha}$ = 0.15-0.4 corresponds to the cooperative conformation transition, and $\overline{\alpha}$ > 0.4 corresponds to the disordered PMAA-S. It should be noted that the titration curve has an inflection (and plateau) in the conformation transition region. For the
network polyacrylic acid sample (an acrylic acid copolymer with 2.5 mol.%
EDMA, curve 3 in Figure 2.8 and called PAA-S), in which the polyelectrolyte
fragments of the network can be imagined to be flexible unstructured chains
by analogy with a soluble homopolymer, the titration curves rise monotonically (compare Figure 2.6b).

The changes in the mobility of the polyelectrolyte fragments in network copolymers during ionization must be valid for every loosely linked
network structure based on MAA because theoretical calculations have shown
that the secondary structure can form when more than 30 monomer links are
present[177]. Thus the formation of the secondary structure from PMAA
fragments in a NPE with small quantities of divinyl crosslinking agent
need not depend on its nature. The titration curves of two carboxylic acid
polyelectrolyte suspensions - MAA with 2.5 mol.% EDMA (Figure 2.9a) and MAA
with 2% divinylbenzene (KB-4-P2) - are given in Figure 2.9.

It is clear from a comparison of the titration curves that the conformation transitions of the NPE samples studied take place in the narrow
band of pH that is characteristic for the disruption of the secondary
structure of the linear PMAA. Increasing the ionic strength of the solution (μ) shifts the titration curves because the ionic strength of the
external solution, created by sodium chloride, affects the acidity of the
ionite. The effect of a neutral salt on the ionization of a cationite[178]
can be described by the empirical equation $pK = pK_0 - \beta \log \mu$, where pK is

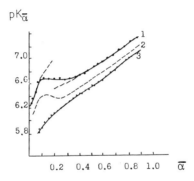

Fig. 2.8. Potentiometric titration curves for copolymer suspensions based on 1) PMAA-S, 2) linear PMAA, and 3) PAA-S.

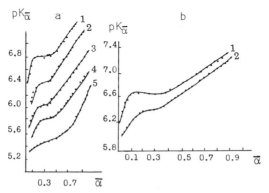

Fig. 2.9. Changes in pK during potentiometric titrations. For suspensions of copolymers of MAA and 2.5 mol.% EDMA (a) and MAA and 2% DVB (b) in 1) 0.001 N NaCl, 2) 0.01 N NaCl, 3) 0.05 N NaCl, 4) 0.15 N NaCl, and 5) 0.5 N NaCl.

the negative logarithm of the apparent ionization constant defined for $\bar{\alpha} = 0.5$, pK_0 is the negative logarithm of the ionite's ionization constant for $\log \mu = 0$, and β is a proportionality constant. Adding electrolytes reduces the mutual repulsions of the ionic charges on the polymer chains as a consequence of screening, i.e., the free energy of the polymer chain electrostatic interactions is reduced, and the structured form of the PMAA is stabilized somewhat.

The salt concentration in the external solution, as can be seen, affects both the $\overline{pK}$ and the pK at which the conformation transition occurs, designated pK_{ct}. Increasing the ionic strength of the solution reduces both pK_{ct} and $\overline{pK}$, while the start of the conformation transition $\bar{\alpha}_{ct}$ is shifted in the direction of greater ionization, i.e., toward greater charge densities on the polyelectrolyte chains. The linear change in pK and pK_{ct} with respect to the logarithm of the salt concentration is clear from Figure 2.10, as is the linear dependence of the shift in the ionization level at which the conformation transition starts as the ionic strength of the solution is increased from 0.01 N NaCl. The linear dependence of $\bar{\alpha}_{ct}$, which corresponds to the start of the conformation transition, on the salt concentration in the external solution also occurs for suspensions of the KB-4-P2 (MAA + DVB) ionite. The conformational transition region (from titration data) is characterized by a constant apparent polyelectrolyte ionization coefficient. For loosely linked network structures based on MAA

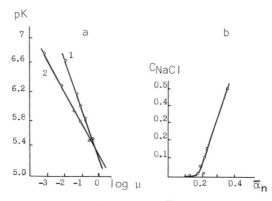

Fig. 2.10. a) Dependence of 1) pK (for $\bar{\alpha}$ = 0.5) and 2) pK_{ct} on the logarithm of solution ionic strength. b) Influence of salt concentration on the start of the conformational transition in PMAA-S suspensions.

the value of pK_{ct} and the range of ionization level at which the conformation transition occurs at a given ionic strength are close (Table 2.1).

If the nature of the titration curves of soluble polyelectrolytes mainly depends on the disposition of the charged groups on the polymer chains, then for the crosslinked analogues the disposition of the chains themselves must be important. The chain disposition depends on the number of chemical bridges and their regularity, which in turn depends on the conditions under which the network structure was formed. Network structures, heterogeneous in density, may be formed during heterophase copolymerization [11,130,181]. The degree of heterogeneous mass distribution within the polymer depends on the concentration of crosslinking agent and the media in which the copolymerization is taking place. The heterogeneity of network structures for MAA and EDMA copolymers grows with increase in crosslink ratio and decrease in the "quality" of the solvents (for example, from 30% acetic acid to pure water). Reducing the solvating power of the solvent (to 5% acetic acid or to water) leads to heterogeneous distributions of polymer mass within the network. Increasing the quantity of crosslinking agent (4 mol.% and 10 mol.% EDMA) can also increase the network's structural heterogeneity. Electron microscopic studies of thin sections of these copolymers[66] have shown that the regions of heterogeneity in an MAA copolymer with 5% EDMA are about 100 Å in size, while those of MAA copolymer with 10% EDMA reach 1000 Å.

Titrating finely dispersed suspensions has shown (Figure 2.11) that for all network MAA copolymers there are, at higher concentrations of crosslinking agent, loosely linked domains. These are shown up, as we have demonstrated, by the presence and disruption upon ionization of secondary structure between the copolymer chains, as this is characteristic of lower density areas or places where the network has been mechanically damaged. All the titration curves have plateaus for ionization levels of $\bar{\alpha}$ = 0.15–0.35, but the shapes of the curves afterwards are, as can be seen from Figure 2.11, anomalous. For network structures obtained from a medium with a good solvent – 30% acetic acid (Figure 2.11, curves 1 and 2) – there is an inflection on the titration curves with a constant pK_{α} (curve 2) for $\bar{\alpha}$ = 0.75–0.85 as the concentration of crosslinking agent is increased from 2.5 mol.% to 4 mol.%. Reducing the "quality" of the solvent (from 30% to 5% acetic acid, curve 3) strengthens this effect, and broadens the region of constant pK_{α} ($\bar{\alpha}$ = 0.65–0.8). When a copolymer with 10% crosslinking agent obtained from a poor solvent is titrated (curve 4), there is a wide band of ionization levels within which pK_{α} does not change ($\bar{\alpha}$ = 0.6–0.9).

34

Table 2.1. Conformational Transitions in Network Polyelectrolytes and Polymethacrylic Acid

Polyelectrolyte	C_{NaCl}	$\overline{pK}$	pK_{ct}	$\overline{\alpha}_{ct}$
PMAA	0.001	6.6	6.4	0.12
MAA + EDMA	0.001	7.0	6.7	0.13
MAA + EDMA	0.01	6.7	6.4	0.18
MAA + DVB	0.001	6.8	6.7	0.12
MAA + DVB	0.01	6.3	6.4	0.18

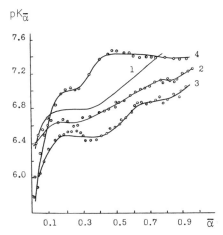

Fig. 2.11. Potentiometric titration curves for MAA-EDMA suspensions. Copolymers obtained in differing concentrations of acetic acid: 1) 2.5 mol.% EDMA, 30% acetic acid; 2) 4 mol.% EDMA, 30% acetic acid; 3) 4 mol.% EDMA, 5% acetic acid; 4) 10 mol.% EDMA, 5% acetic acid.

The dense heterogeneous sections are destroyed by being ionized when $\overline{\alpha}$ is greater than 0.6. They must in fact be considered to be agglomerations of PMAA chains fixed by crosslinking. The high local concentration of structured PMAA chains formed during synthesis and linked into a dense polymer mass may lead to the formation of a set of interchain contacts. This crosslinked structure can only be disrupted at high ionizations. This shows up in the titration curves as the section of constant $pK_{\overline{\alpha}}$ at $\overline{\alpha} > 0.6$, at which the strength of the electrostatic repulsions becomes commensurate with the stabilization energy of the internal structure. The shift of the conformation transition to higher ionization levels is observed, as has been seen, when the ionic strength of the solution is increased for soluble PMAA and PMAA-S.

Potentiometric titration data can also be given as pH versus $\log \dfrac{1 - \overline{\alpha}}{\overline{\alpha}}$ plots. The results of titrations of samples of AKA copolymer with 2.5 mol.% EDMA (PAA-S) and PMAA-S are given in Figure 2.12a. All the points on the PAA-S curve (2) lie on a straight line; this is an example of the titration of flexible unstructured polymer fragments. For PMAA-S (2.5 mol.% EDMA, curve 1) in a Henderson-Hasselbach relation there are a series of straight line segments with different slopes, each corresponding to another stage in the ionization process. The region $\overline{\alpha} = 0-0.15$ corresponds to the titration of PMAA chains in the structured conformation;

the conformation changes in the $\bar{\alpha}$ = 0.15–0.35 region as the secondary
structure is disrupted; the rest of the curve refers to the titration of
the unstructured conformation of PMAA-S fragments. The titration results
for samples of the heteronet copolymer of MAA with EDMA (10 mol.% or 5%
acetic acid, curve 3) are presented in Figure 2.12b. Besides the sections
that were seen for the PMAA-S in Figure 2.12a (2.5 mol.% EDMA), there is a
break in the straight line starting at $\bar{\alpha}$ ∿ 0.6, which is the region in
which the denser formations are presumed to be titrated. For comparison
the results of titrating a suspension of a carboxylic NPE – an MAA copolymer with 12 mol.% hexahydro-1,3,5-triacryloyltriazine (HTA) – is presented
here (curve 4). This has a clearly heterogeneous structure owing to its
polymerization conditions (water solvent) and the parameters of the cross-
linker[143]. Two sections with differing slopes are clearly visible in
Figure 2.12 (curve 4). They correspond to the ionization of the first of
the easily accessible carboxylic groups and then to that of the groups in
the densely crosslinked regions at $\bar{\alpha}$ > 0.6. It seems that the degree of
macro-heterogeneity in carboxylic acid heteronet polyelectrolytes can be
estimated from the titration curves. Based on Figure 2.11 an MAA copolymer
with 2.5 mol.% EDMA (30% acetic acid) is a heteronet structure, while about
10% of the titratable groups of an MAA copolymer with 4 mol.% EDMA (30%
acetic acid) are in the heterogeneous dense sections. When an MAA copolymer with 4 mol.% EDMA is obtained from 5% acetic acid, about 20% of
the titratable groups are found in the heterogeneous section, while the
fraction in the heterogeneous dense section is about 50% for MAA samples
with 10 mol.% EDMA (5% acetic acid)[184].

The degree of heterogeneity of network structures is fixed by the
synthetic conditions. Thus a suspension polymerization in a hydrophobic
medium yields a microgranulated NPE. Compared with a copolymer made as a
slab from the same initial reaction mixture, microgranules can only swell
to a restricted degree when ionized (Figure 2.13). This is due to the more
rigid surface layer of a granule, which hinders the increase in dimension
compared to mechanically ground polymer prepared as a slab. Because the
fixing of polymer chain and functional group arrangement by the crosslinking agent is important for structure formation, it may be assumed that
the rigidity of surface layers is due to large local densities of the
polyelectrolyte chains and their crosslinking at the bead surfaces during
the polymerization.

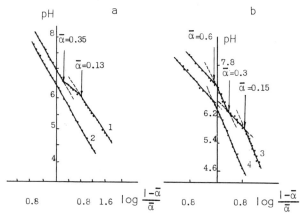

Fig. 2.12. Potentiometric titration of carboxylic cationite suspensions
in Henderson-Hasselbach coordinates. a: 1) PMAA-S and 2) PAA-S
copolymers (2.5 mol.% EDMA) obtained in 30% acetic acid; b: 3)
heteronet copolymer of MAA and EDMA (10 mol.%, 5% acetic acid);
4) copolymer of MAA and HTA (12 mol.%, water).

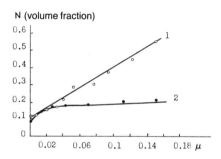

N (volume fraction)

Fig. 2.13. Swelling of MAA and EDMA copolymers versus ionic strength.
            The samples were expanded in a phosphate buffer solution pH 7
            and were obtained under different synthesis conditions. Curves
            1) is for a polymer sample synthesized in a slab and 2) for a
            sample obtained by the suspension method, and both copolymer-
            ized in the presence of 30% acetic acid. Bead size was
            5-10 μm.

    Two types of suspension microgranule copolymers were investigated
experimentally[185]. The microgranule sample of the first type was fractionated (non-destructively), and a suspension with 1-μm particles was
used, and designated PMAA-S-MG. Note that in the finely dispersed microgranule samples (we are dealing with micrometer-sized particles) the
surface may play a considerable role in their properties. In the second
variant the larger microgranules were ground down mechanically and the
fraction with micrometer size was separated, while carefully washing away
the low-molecular-weight components that can arise when the sample is
broken down. The rigidity of carboxylic acid polyelectrolyte systems can
be compared with respect to their swelling coefficients in the ionized
($\bar{\alpha} = 0.5$) and un-ionized ($\bar{\alpha} = 0$) forms. Table 2.2 contains data on the
swelling of fractions of copolymer prepared as a slab, of microgranule
samples, and of ground microgranules. It can be seen from the table that
the method of formation and the mechanical grinding have practically no
effect on the swelling of the electrolytes in the un-ionized states. There
is a considerable difference on ionization; the grinding of the microgranules leads to an increase (almost double) in the volume of the network
at $\bar{\alpha} = 0.5$.

    The potentiometric titration data are presented in Figure 2.14 for
suspensions of microgranule and ground microgranule samples. In the
PMAA-S-MG sample homogeneous carboxylic acid groups are titrated from
$\bar{\alpha} = 0$ to $\bar{\alpha} \sim 0.6$. The break in the curve in the region $\bar{\alpha} = 0.6-0.76$ shows
another ionization regime of the carboxylic acid groups. This may be
ascribed to the titration of local sections of rigid microgranule surface
structures where high concentrations of carboxylic acid groups have been
formed that hinder ionization[186]. The internal structure (fixed in by
crosslinking) of local densely packed chains of the PMAA-S-MG sample can
only be disrupted at high ionization levels. The electrostatic repulsions
between like charges on the chains can then break the bonds stabilizing the
structure. This is similar to the situation we presented earlier for
heteronet structures which either had large crosslink ratios or had been
obtained under conditions promoting heterogeneity[184].

    The titration of the ground microgranule samples (Figure 2.14, curve
2) shows up the two types of ionization in the network structure. The
first can be ascribed to the titration of the interiors of the granules and
those sections mechanically damaged when the network was broken down, and
the mobility of the PMAA chains improved. As was the case with loosely
linked macronet structures, the PMAA fragments behave like linear PMAA when

37

Table. 2.2. Relationship between the Swelling Factors of Network
Polyelectrolytes

NPE	$K_{sw}(\bar{\alpha} = 0)$	$K_{sw}(\bar{\alpha} = 0.5)$	$\dfrac{K_{sw}(\bar{\alpha} = 0.5)}{K_{sw}(\bar{\alpha} = 0)}$
1. MAA + 2.5 mol.% EDMA in the presence of 30% acetic acid prepared as a slab	4.5	11.5	2.7
2. MAA + 2.5 mol.% EDMA with 30% acetic acid, obtained by suspension polymerization	4.0	8.8	2.2
3. Ground sample of (2)	4.5	17.0	3.8

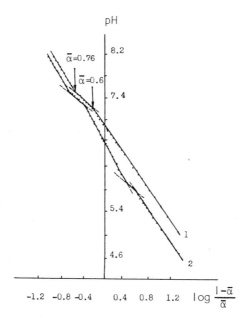

Fig. 2.14. Potentiometric titrations of network carboxylic acid
polyelectrolytes. 1) PMAA-S-MG (polymethacrylate network
microgranules); 2) PMAA-S-MG, ground sample.

ionized, i.e., they have a secondary structure at low ionizations which is
disrupted at $\bar{\alpha} = 0.15-0.3$. The second type of ionization is of the denser
sections, which as can be seen from Figure 2.14 are able to ionize, like
heteronets, with the disruption of the interchain contacts at $\bar{\alpha} = 0.7-0.8$.

It is known[187] that the free electrostatic energy of most ionites is
due to chain interactions and is proportional to $\log\dfrac{\bar{\alpha}}{1-\bar{\alpha}}$ for a wide band
of ionization levels of polymer acids (from $\bar{\alpha} \sim 0.1$ to $\bar{\alpha} \sim 0.9$). In the
empirical equation for this relationship, viz., $pH = pK_{typ} - n\dfrac{1-\bar{\alpha}}{\bar{\alpha}}$, the
parameter n reflects the electrostatic repulsion of charged polyelectrolyte
chains. Thus the deviation of n from 1 is a measure of the electrostatic

38

effect. It can be determined from the slope of the linear section of a titration curve plotted in pH versus $\log \frac{1 - \bar{\alpha}}{\bar{\alpha}}$ coordinates.

The n parameter for a NPE characterizes the influence the polymer chains within the network have on one other. Differences in the interactions between the polyelectrolyte fragments at different ionizations of the macronet loosely knit MAA copolymer with 2.5 mol.% EDMA show up in the different n for the structured (I) and unstructured (II) forms in the titration of PMAA-S (see Figure 2.12a). Changes in the n parameter during the ionization of the NPE in question are given in Table 2.3. When PMAA-S fragments having secondary structure (I) are titrated, the ionization region ($\bar{\alpha}$ = 0.15-0.35) in which the ordered structure is disrupted encompasses the usual conformation transition. Then comes the titration of the loosely bound PMAA coils between chemical crosslinks, and finally the ionization range $\bar{\alpha}$ = 0.6-0.8 in which the areas more dense and more difficult to neutralize are eventually ionized.

It follows from Table 2.3 that n for the usual conformational transition region is close to unity in all cases. This means that the dis-

Table 2.3. Changes in n during the Ionization of a NPE

				n		
					Internal	
		I	Conform. transition	II	structure destroyed	
Sample	μ	$\bar{\alpha}$ = 0.0- -0.15	$\bar{\alpha}$ = 0.15- -0.4	$\bar{\alpha}$ > 0.4	$\bar{\alpha}$ = 0.6- -0.8	$\bar{\alpha}$ > 0.8
			Macronets		Heteronets	
PMAA	0.001	1.5	1.0	1.74	–	–
PMAA–EDMA	0.001	1.46	1.0	1.68	–	–
(2.5 mol.%)	0.01	1.48	1.15	1.67	–	–
	0.10	1.40	0.90	1.75	–	–
	0.15	1.45	1.17	1.66	–	–
PMAA–EDMA (2.5 mol.%) microgranules	0.001	1.40	–	–	1.14	1.40
PMAA–EDMA (2.5 mol.%) ground microgranules	0.001	1.44	1.0	1.75	1.12	1.40
KB-4P-2 (MAA-2% DVB) ground sample	0.001	1.56	0.90	1.52	–	–
	0.01	1.65	1.0	1.69	–	–
PMAA–EDMA (4 mol.%) 5% CH₃COOH	0.001	1.56	0.6	1.74	0.8	1.4
PMAA–EDMA (10 mol.%) 5% CH₃COOH	0.001	2.50	1.0	1.80	0.9	–
MAA–HTA (12 mol.%)	0.001	2.90	–	–	1.16	–

ruption of the secondary structure of PMAA only occurs when the increase in charge density during ionization is compensated by the decrease in the energies of the interaction between the chains that results from the disruption to the stabilizing structure of the bonds. The value of n becomes close to unity even for sections, during a heteronet titration, that belong, as we said before (at $\bar{\alpha} > 0.6$), to the anomalous ionization of the denser areas.

As we have mentioned, the formation and disruption of a structure is above all a reflection of the intramolecular mobility of polymer chains (Figure 2.6). Changes in the relaxation time during the ionization of PMAA-S and PMAA-S-MG are given in Fig. 2.15. The mobility of the chains in the microgranular PMAA-S-MG is strongly hindered and much lower than that in PMAA-S. The greatest mobility in PMAA-S-MG occurs at high ionizations $\bar{\alpha} \sim 0.7$ (for PMAA-S $\bar{\alpha} \sim 0.4$). We can assume that the internal structure is retained in the PMAA-S-MG until high ionizations, which is evidence of the strong intermolecular contacts in the dense areas, probably, of the heterogeneous structure[188]. A comparison of the relaxation times τ_w of network and linear polyelectrolytes at different ionizations shows that the cross-links and conditions of network formation strongly influence the dynamic parameters of the polymer fragments (Table 2.4). These data are evidence of stronger interchain contacts between linear fragments in network polyelectrolytes (obtained in different conditions) than in linear structures, and they agree with the results of potentiometric titrations.

In this way the use of finely dispersed suspensions of carboxylic acid NPE's based on MAA has enabled potentiometric titration and polarized luminescence to be applied, together with data on changes in volumetric swelling, to the study of the structure and dynamic behavior of network systems. The papers we have considered here were the first to note and corroborate quantitatively the changes in the mobility of polymer chains

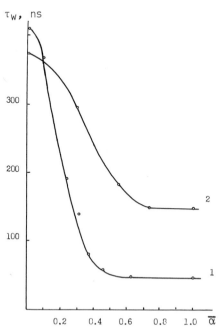

Fig. 2.15. Changes in intramolecular mobility during ionization. For MAA-EDMA (2.5 mol.%) copolymers prepared in the presence of 30% acetic acid 1) in slab form and 2) by the suspension method.

Table 2.4. Changes in the Intramolecular Mobility of Linear Fragments of Network Structures for Different Ionizations

NPE*	$\tau_{net}/\tau_{linear\ PE}$		
	$\bar{\alpha} = 0$	$\bar{\alpha} = 0.5$	$\bar{\alpha} = 1.0$
PAA–EDMA (2.5 mol.%) in slab form	1.6	2.0	2.0
PMAA–EDMA (2.5 mol.%) in slab form	4.5	4.0	4.0
PMAA–EDMA (2.5 mol.%) in microgranule form	4.2	15.0	12.0

*NPE synthesized in the presence of 30% acetic acid.

between the crosslinks during ionization. The character of these changes has been shown to be entirely determined by structural features of a network polyelectrolyte. The mobility of the polymer chains in heteronet structures depends on whether the chains can be freely arranged between the crosslinks (then they behave like linear PMAA) or whether they are effectively fixed by the crosslinking agent into the dense sections, in which case their mobility is very limited.

It must be assumed that the structural and dynamic characteristics and conformational states of the chains, as well as the changes in the mobility of the linear fragments in a NPE, all have a substantial influence on the network's interaction with organic ions, especially with complicated ones, e.g., proteins.

3
Ion-Exchange Equilibrium, Thermodynamics, and Sorption Selectivity of Organic and Physiologically Active Substances

Selective interaction with ionites is at the heart of sorption ion-exchange separation of organic and particularly physiologically active substances. This is true not only for static sorption, but also for the dynamic frontal processes. When constructing a theory for the equilibrium dynamics of ion-exchange sorption, the selectivity constants and sorption limiting capacity are major factors in the criteria defining the efficiency of a process. Finally, even in non-equilibrium dynamic sorption, selectivity, together with kinetic and hydrodynamic parameters and the dimensions of the columns, is an important indicator of i) whether a substance can be separated, ii) the yield on desorption, and iii) the concentration in the eluant. When considering the ion-exchange sorption of organic ions, especially complex ones like those of physiologically active substances, we must take into account the size of the ions, the complex polyfunctional interactions with the ionite, the morphological and electrochemical properties of the ionite, and other features of the ionites, the counter-ions, and the liquid medium in which the heterogeneous ion-exchange occurs.

A diagram demonstrating the basic components of a polymer network polyelectrolyte (cationite) is given in Figure 3.1. The polymer skeleton is sometimes called the ionite's matrix. The linear polymer chains are connected in network structures by bridging groups (crosslinks). Ionogenic groups in the polymer chains, covalently bonded to the matrix, are like fixed ions. Counter-ions are mobile and can be exchanged for another ion with the same sign in the solution. A fixed ion with a counter-ion can be considered as an ionogenic group of the sorbent. Moreover, the concept of an ionogenic group may also be applied to a weak electrolyte when there is an un-ionized ion pair instead of a fixed-ion—counter-ion system. When an electrolyte is absorbed from a solution with a high ionic strength (like the Donnan effect), more counter-ions enter the ionite than there are fixed ions, and this leads to the sorption of co-ions with the same sign as the fixed ions. The solvent (usually water) in the ionite solvates each sort of ion and may also interact weakly with the ionite's structural elements. Weak interaction corresponds to the "free" solvent ("free" water) state. This scheme does not allow for some of the features of ion-exchange materials such as the formation of ions pairs for weakly solvated insoluble electrolytes, complicated polyfunctional (additional) interactions of organic ions, the limited accessibility of some sections of the ionite to large ions, and special ionite types (e.g., pellicular, surface-layer, and bidispersed ionites). We shall consider these aspects later not only in

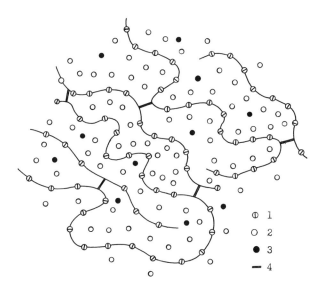

Fig. 3.1. Schematic ionite structure. 1) Fixed ion; 2) counter-ion;
3) co-ions; 4) crosslinks or bridging groups from crosslinking agent.

terms of schemes but also by strict descriptions of the structures and
their associated laws.

3.1. IONITE EXCHANGE CAPACITY

If the sorption capacity of an ionite is understood to be the number
of ions it can absorb from a solution under any experimental conditions
and in competition with other counter-ions, then its exchange capacity is
conventionally taken to be its limiting absorption ability. Only interactions between counter-ions and fixed groups are considered here, while
Donnan absorption of electrolytes and non-exchange sorption, which is
especially possible for organic electrolytes and is associated with the
appearance of polymolecular layers in the canals and on ionite granules,
particularly during flocculation, are excluded. Thus the most important
determinant of exchange capacity is the number of fixed ions or ionogenic
groups in the ionite, although the number of counter-ions may differ from
that of the fixed ions. Direct empirical chemical analysis very rarely
gives a true picture of the number of ionogenic groups. For example, the
sulfur in a sulfocationite may occur in sulfate bridges, and hence the
exchange capacity calculated from the sulfur content will be greater than
the actual exchange capacity[54]. The number of counter-ions associated
with the fixed groups in an ionite, i.e., its exchange volume, depends on
the medium's pH, which determines the ionization of the ionogenic groups.
This ionization depends not only on the acidity or basicity of the groups
but also on the influence of neighboring ionized groups.

One way of measuring the number of associated counter-ions is potentiometric titration. It is applied to weighed samples of ionite[65,71]
with different quantities of acid or alkali (usually 0.1 N). After the
equilibrium has been established, which takes 1-5 days, the solution's pH
is found. The best form for the titration curves for our purposes are
those depicting the relationship between solution pH and the milligram
equivalent of the alkali or acid added to the sample per unit weight of
dehydrated ionite. The curves for a sulfocationite, a carboxylic cationite, a phosphoric cationite, and an anionite are given in Figure 3.2.

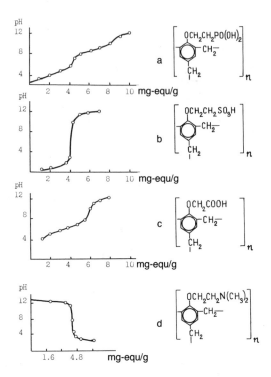

Fig. 3.2. Potentiometric titration curves for ionites a) KFE, b) KFS,
c) CPA, and d) ADG.

We see that the polycondensed ("condensation polymerized") sulfocationite
KFS has a constant exchange capacity (number of sorbed sodium ions),
approximately 4 mg-equ/g over a wide range of pH. It is easy to detect two
titration regions for phosphocationites, one for each of the two acidic
protons attached to the phosphorus ion. The cationite's exchange capacity
reaches 8 mg-equ/g at pH 10. For polycondensed anionites the exchange
capacity is close to 4 mg-equ/g. The exchange capacity of a polycondensed
carboxylic cationite grows slowly as the solution's pH is increased,
reaching 5-6 mg-equ/g. The ionization of polyelectrolytes (and indeed for
all electrolytes with closely placed ionogenic groups) is unusual in that
as the ionization proceeds the energy of interaction between counter-ions
and fixed ions grows because of the increase in the electric potential of
the ionite matrix (interactions not only take place with near fixed ions
but also with its neighbors), and this leads to the ionogenic groups being
less ionizable. Polymer electrolytes are thus weaker electrolytes than
their corresponding monomers. The most useful theory for the potentiometric titration of polyelectrolytes is the one developed by Katchalsky and
his colleagues[120,133-135]. It describes a polyelectrolyte's ionization
by the equation

$$pH = pK_{typ} - \ln \frac{1 - \bar{\alpha}}{\bar{\alpha}} + \frac{F\psi_0}{2.3RT}, \qquad (3.1)$$

where $\bar{\alpha}$ is the polyelectrolyte's ionization level,
ψ_0 is the polyanion's electrostatic potential, and
F is Faraday's number.

When studying ionization, especially for densely crosslinked poly-electrolytes (polyacids), Gregor's empirical equation is often used, i.e.,

$$pH = pK - n \log \frac{1 - \bar{\alpha}}{\bar{\alpha}}, \tag{3.2}$$

where n is a constant that depends on all the types of interaction in the electrolyte, including the electric potential of the matrix, the conformational entropy of the polymer network, and interactions between chain sections lying between both physical and chemical crosslinks. The potentiometric titration curves given in Figure 3.3 use Henderson-Hasselbach coordinates and were obtained for carboxylic cationites. They demonstrate the rise in acidity – ionization – as the ionic strength of the solution is increased, which corresponds to ion screening and the reduction of the polymer's total potential with respect to counter-ions.

An ionite's exchange capacity depends on the type of counter-ion. For example, carboxylic cationites sorb divalent metal ions in preference to monovalent ones. Moreover, both thermodynamic selectivity and exchange capacity contribute to this. In order to appraise the effect pH has on the exchange capacity for divalent metal ions the following equation is used[144]:

$$pH = pK - n \log \frac{1 - \bar{X}_m}{\bar{X}_m} \tag{3.3}$$

which is similar to Gregor's equation, and where $\bar{X}_m$ is the equivalent fraction of the divalent ion in the ionite.

An analysis[12,144] of the absorption of divalent ions by carboxylic cationites has shown that the pK for these systems is less than the pK_{typ} of sodium when it is absorbed by the same ionites. This indicates that the exchange capacity of the carboxylic cationites is large for divalent ions at a given pH.

The degree of crosslinking in an ionite largely influences its exchange capacity, but there are other factors depending on the ionite and counter-ion. A common reason for a fall in exchange capacity for nearly all ionites and counter-ions is the reduction in the number of ionogenic groups per unit mass of ionite when the crosslink ratio is increased, because crosslinks do not usually contain ionogenic groups. For example, an increase in the concentration of divinylbenzene (DVB), in the sulfonated copolymer of DVB and styrene, from 1% to 20% reduces the exchange capacity by about 20%. There is a second factor primarily influencing weak network polyelectrolytes. An increase in the volumetric concentration of the

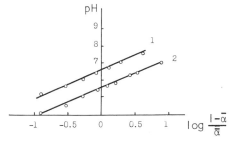

Fig. 3.3. Potentiometric titration of KM-2P carboxylic cationite for different ionic strengths: 1) 0.1 N; 2) 1 N.

ionogenic groups in the denser sections of the polymers leads, according
to the principles we have presented, to an increase in the free energy of
the electrostatic interaction with counter-ions, and consequently to a
reduction in the ionization of weak polyelectrolytes. The way in which
the ionization levels of a macroporous carboxylic cationite depend on the
ionite's degree of crosslinking is shown in Figure 3.4.

The degree of crosslinking is the predominant factor in the exchange
capacity of an ionite with respect to organic counter-ions. A large factor
here is the inaccessibility of some of the ionogenic groups due to dense
parts of the net hindering the organic ions entering the ionite from the
solution. If we exclude kinetic factors, i.e., we overcome what we shall
call the ionite's kinetic permeability, then the limiting absorption
capacity of an ionite with respect to organic ions turns out to be
dependent on the ionite's degree of crosslinking and the molecular mass of
the ion. Figure 3.5 shows how the exchange capacity of a sulfocationite
SBS, which is a styrene and divinyl copolymer, depends on its volumetric
swelling, which is an inverse function of the percentage content of crosslinking agent. Cationites with swelling factors of 1.2 have an exchange
capacity with respect to tetracycline that is only 1% that with respect to
mineral ions (H-Na exchange). When the swelling factor is 6.0, the
exchange capacity with respect to tetracycline approaches that of
sodium[145]. It is useful here to introduce the idea of relative exchange
capacity of ionites for organic counter-ions in the form

$$N = \frac{M100}{m} \, , \qquad\qquad (3.4)$$

where m is the exchange capacity with respect to small ions, and
M is the exchange capacity with respect to the organic ion.

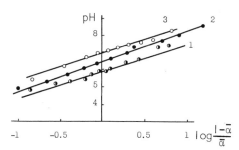

Fig. 3.4. Potentiometric titration of SGK-7 carboxylic cationite for
different crosslink ratios: 1) 10%; 2) 15%; 3) 30%
DVB; 8°C; ionic strength 0.1 M.

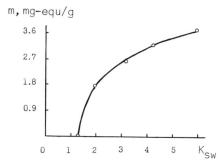

Fig. 3.5. Exchange capacity of SBS sulfocationite for tetracycline
versus swelling factor.

For a number of organic ions their relative exchange capacities can
be increased if the number of ionogenic groups is decreased for the same
swelling factor. Thus in going from the polycondensed cationite CPA, which
is obtained by condensing phenoxyacetic acid with formaldehyde, to the
cationite CPAC, which is condensed from phenoxyacetic acid, p-chloro-
phenol, and formaldehyde, the relative exchange capacity with respect to
the antibiotic streptomycin grows from 50% to 100%, although the swelling
factor is changed slightly and the exchange capacity with respect to sodium
is diminished from 5.6. mg-equ/g to 2.8 mg-equ/g[57]. A similar situation
is observed when anionites sorb nucleotides (Figure 3.6). The relative
exchange capacity with respect to adenosine monophosphate and uridine
monophosphate on Dowex-1 × 2 type anionites is reasonably high only after
the partial deamination of the ionite, during which the number of ionogenic
groups is reduced 5-10-fold. This can be interpreted as the covering of
several ionogenic groups by one organic counter-ion that can bond only to
one fixed ion. As a result, the neighboring group or groups are rendered
inaccessible to organic counter-ions. Reducing the exchange capacity of an
ionite does not alter the relative exchange capacity for small organic ions
such as triethylbenzylammonium (TEBA). However, in these systems it is
useful to note the changes in the ionization of a weak polyelectrolyte
(carboxylic cationite) when titrated with caustic soda and a solution of
the base TEBA-OH (Table 3.1).

It can be seen from Table 3.1 that introducing ions of the organic
base increases pK_{typ}, i.e., reduces the ionization of a carboxylic cationite, bringing down at the same time the exchange capacity with respect
to organic ions at a given solution pH. Interestingly when hydroxylpro-
pylmethacrylamide is added to the copolymer and the volumetric concentration of the carboxylic groups is reduced, the impact of the polymer
effect on pK_{typ} is only observed when the concentration of the methacrylic
acid monomer in the copolymer starts to grow. At the same time, the parameter n, which according to Gregor characterizes the deviation in the
polyelectrolyte's behavior from that of the monomer acid (for which n = 1),
is only apparent for large volumetric concentrations of the carboxylic
groups. Thus, changing the volumetric concentration of the carboxylic
groups by using non-ionogenic monomers affects the degree to which carboxylic cationites ionize in a more complicated manner than when the crosslink ratio is increased[16]. The concept of relative exchange capacity
must be used carefully because of the "pseudo-equilibria" that can occur in
these systems[12]. A large number of papers have demonstrated that the
permeability of an ionite for counter-ions grows as the temperature is increased, as the sorbent is ground down, or as the solution's ionic strength

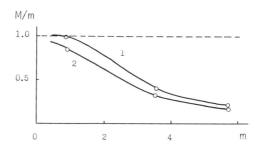

Fig. 3.6. Relative exchange capacity of Dowex-1 anionite with respect to
1) adenosine monophosphate and 2) uridine monophosphate versus
ionite deamination. m is the anionite exchange capacity for
small ions; M is the exchange capacity of the anionite for AMP
and UMP.

Table 3.1. Potentiometric Titration of the Network Copolymer of Methacrylic Acid, Hydroxypropyl Methacrylate, and Hexahydro-
1,3,5-triacryloyltriazine

Concentration of carboxylic groups in the ionite		NaOH Titration		TEBA-OH Titration	
mg-equ/g	mg-equ/ml	pK_{typ}	n	pK_{typ}	n
1.89	0.18	6.3	1.0	6.8	1.0
3.56	0.22	6.9	1.1	7.3	1.0
6.30	0.39	6.9	1.0	7.9	1.3
7.33	0.89	7.0	1.35	7.9	1.4

is increased. These changes may also alter somewhat the relative exchange
capacity, too, due to changes in the permeability and the accessibility of
the active centers. At the same time, we must remember all the inter-
related changes that occur in a network polyelectrolyte's properties when
the temperature, bead dispersity, or ionic strength are changed, or when
organic ions are introduced, etc. We have shown that introducing organic
ions alters the ionization level in connection with changes in the
ionite's dielectric constant. This cannot be related to the question of
permeability but does arise for changes in relative exchange capacity.
Furthermore, when heteronet ionites are ground up, the less-dense sections
are damaged, and this entails changes in the electrochemical properties of
the ionite because the densely crosslinked and open sections are characterized by different potentiometric titration constants. All this indicates that the complex interrelations between the possible changes in an
ionite's properties cannot simply be considered as part of the question of
permeability and accessibility of the functional groups to counter-ions.
Meanwhile there are some ionite types whose relative exchange capacities do
not change, or alter only slightly, when the parameters we are considering
are changed. This is true of the macroporous ionites, in which there is
a clear differentiation between the dense network sections and the open
spaces - the pores or canals. In these ionites only the spaces are accessible to large organic ions, such as the insulin macromolecule, and
sorption occurs only on the canal walls[149]. The dense regions and the
functional groups within them are inaccessible to insulin whatever the
conditions.

A variation in the relative exchange capacity thus depends on many of
the system's parameters and may be significant or negligible. Because of
the significance of this parameter, we should look at it with respect to
both changes in an ionite's permeability and to alterations in its structure and electrochemical properties given different external agents. For a
theoretical analysis, the relative exchange capacity must remain constant
and hence some of the parameters may only vary within limits.

The sorption capacity (including the exchange capacity) of an ionite
with respect to an organic ion may be significantly changed when the ionite
is ground down[159]. The change is particularly great for very large
organic ions, such as protein macromolecules (Figure 3.7). When using
heteronet ionite beads 1-3 μm in diameter, the sorption capacity with
respect to serum albumin and hemoglobin reaches the astonishing value of
5-10 g/g[179]. There is a significant growth in sorption capacity even for
30- to 40-μm fragments when the ionite is ground down. A feature of the
absorption of proteins by microdispersed ionite beads is the existence of
a maximum on the plot of sorption capacity versus bead dimension. Beads

48

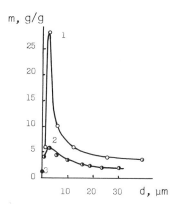

Fig. 3.7. Sorption capacity for blood albumin by sulfocationites versus ionite particle size. 1) KU-23; 2) Dowex 50 × 1; 3) polystyrene; pH 1.65.

with diameters less than a micrometer sorb less protein than do beads 1-3 μm in diameter. The small absorptive capacity of ionites at high dispersities may be compared with the limited ability of soluble polyelectrolytes to bond with protein macromolecules. It would seem that the known reduction of specificity of intermolecular interactions is being observed here[180]. The fall in the sorption capacity with respect to proteins at high ionite dispersities is apparently not a consequence of limiting absorption, which we are studying, but of an absorption level that depends on the concentration of the protein in the solution. This is because the sorption of proteins, like that of simple dipolar ions, is not the classical exchange of equivalent ions and the limiting sorption capacity - exchange capacity - only occurs at high protein (and amino acid and peptide) concentrations. This occurs even in the absence of competing ions. This phenomenon will be discussed further later.

Besides the static methods of determining exchange capacity, dynamic methods are also used. An electrolyte solution is passed through a column containing an ionite, and the counter-ions in the ionite are exchanged for the ions with the same sign that are fed in. The final state defines the equilibrium with the feed solution and the desorbed ions within the column. Thus, given the introduction of one type of counter-ion, a dynamic process may be used to determine the number of ions that can be replaced by the counter-ions under the test conditions (for the given pH, solvent, and temperature). Thus a dynamic experiment yields the exchange capacity with respect to the feed counter-ion. The quantity may be calculated with respect to either the number of sorbed or desorbed ions given equivalent ion exchange.

3.2. EQUIVALENCE OF ION EXCHANGE

The chemical equivalent of the sorbed ions is in most cases of ion exchange equal to the chemical equivalent of the desorbed ions. This may be traced in both static and dynamic experiments. However, equivalence is often violated, particularly when certain organic ions participate in the ion exchange. As can be seen from Figure 3.8, when the base oleandomycin is sorbed in a dynamic regime onto a carboxylic cationite from a neutral solution, the amount of desorbed sodium exceeds the amount of sorbed organic ion. This means that the sodium counter-ions are replaced not just by the organic ions but also by hydrogen ions, and this is verified by a shift in the pH of the solution discharged from the column. Thus the H-Na

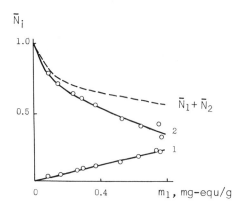

Fig. 3.8. Mole fraction of oleandomycin and sodium counter-ions in KRFU
carboxylic ionite for differing degrees of loading of the
antibiotic. m_1 is the capacity of the ionite for oleandomycin.
Curve 1 is for oleandomycin; curve 2 is for sodium.

equilibrium in a carboxylic ionite is also dependent on the presence of
organic counter-ions and is determined by the dielectric constant of the
media, and may be compared with the effect of the organic solvent feed.
The ionization of the carboxylic cationite also falls. Thus the difference
in exchange capacity between organic ions and mineral ions not only depends
on the reduced accessibility of the ionogenic groups to the larger ions,
but also on the change in the ionite's ionization. The differences between
the potentiometric titration curves for carboxylic ionites by caustic soda
and those by organic bases (Figure 3.9) should be interpreted in the same
way.

In a number of systems with organic sorbates the charge on the
counter-ion is increased when it goes from the solution to the ionite.
This occurs, for example, when uridine monophosphate is sorbed onto
Dowex 1 × 2 anionite (Figure 3.10) [182]. The change in ionization towards
increased counter-ion charge is most clearly seen from the sorption of
α-amino acids[183]. The dipolar ions of α-amino acids carry, in neutral
solutions, both a positive and a negative charge, which correspond to the
ionization of the amine and carboxylic groups. However, when they transfer
to the hydrogen form of a sulfocationite, the dipolar ions of glycine,
alanine, and other amino acids eject the hydrogen ions into the solution
(Figure 3.11), which corresponds to

$$RSO_3^-H^+ + H_3\overset{+}{N}RCOO^- \rightleftarrows RSO_3^-H_3\overset{+}{N}RCOOH.$$

Thus the investigator observes the absence of equivalence. It should be
noted that as the distance between the positive and negative charges
increases, this effect degenerates, and the ion exchange of many ω-amino
acids reverts to the classical form. Thus the longer dipolar ion interacts
with the fixed ion of the ionite to eject the matrix counter-ion into the
solution, while the other end retains its charge. A reason often advanced
for the change in the charge of α-amino acids is the high acidity of the
media in beads of the hydrogen form of a sulfocationite. However, this is
only one reason for the non-equivalence or apparent non-equivalence of ion
exchange because it has been shown that the amino acids are sorbed using
the same mechanism at high concentrations too, that is, when all the ionogenic groups are occupied by the organic counter-ions and the local concentrations of hydrogen ions are very small. Moreover, the sorption of
dipolar amino-acid ions without the ejection of their chemical equivalent
of hydrogen counter-ions not only occurs on sulfocationites, but also on

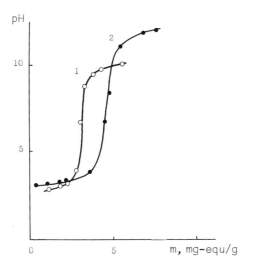

Fig. 3.9. Potentiometric titration curves for Dowex 50 × 1 by 1) oleandomycin base and 2) NaOH; ionic strength 0.003.

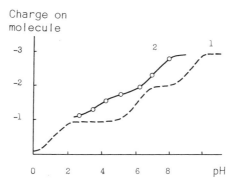

Fig. 3.10. Equivalence of exchange of uridine monophosphate on Dowex-1
(Cl form) for differing pH's. 1) Charge on UMP in solution;
2) charge on UMP sorbed on anionite.

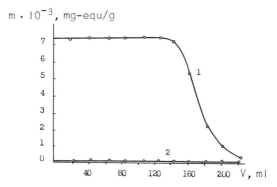

Fig. 3.11. Absence of equivalence in the sorption of alanine on SDV-3
sulfocationite. Curves 1) alanine; 2) hydrogen. pH at exit
of column did not change.

ionites that contain phosphonic acid residues. The effects we have described are more easily seen in dynamic column experiments, in which it is
possible to avoid the acidification of the external solution and to observe
the equilibrium with the initial solution. Experimental data recently
gathered indicate that the ionization states of both the fixed ions in the
ionite and the organic counter-ions are functions not only of acidity,
local acidity in particular, but also of the dielectric properties of
environment of the ionite beads, which is determined by both the ionite
matrix and the properties of the counter-ions and their concentration in
the sorbed states.

3.3.   ION-EXCHANGE PROCESSES WITHOUT THE SORPTION OF PHYSIOLOGICALLY ACTIVE SUBSTANCES

Mineral ions and small ions in general are sorbed onto ionites from
solutions containing physiologically active substances (particularly those
that are electrolytes). This situation is important in stages involving
the neutralization, demineralization, or counter-ion substitution of electrolytes such as antibiotics, alkaloids, and proteins.

The preparative substitution of counter-ions in polyelectrolytes is
very simply accomplished by ion exchange in a column in a dynamic regime.
Thus in order to convert the sodium salt of benzylpenicillin to the potassium salt (or vice versa) the solution needs only to be filtered through a
cationite solution[121]. However, the apparent simplicity of the process
is complicated by the significant loss of penicillin due to its sorption by
synthetic organic cationites with the formation of hydrogen bonds[122]. A
full yield is obtained by passing the antibiotic solution through an
aluminosilicate-permutite cationite (Figure 3.12). However, this only has
practical value for the conversion of the potassium salt of penicillin to
the sodium salt, because if the sodium salt is passed over the potassium
form of the permutite at rapid solution flow rates (100 ml/cm^2·h), sodium
ion overshooting sets in very quickly. We note here, in advance of our
detailed presentation of the topic, that for useful substitutions in
ion-exchange columns there must be conditions for the formation of sharp
boundaries between the ion zones. For slow solution flow rates the formation of sharp boundaries in the case of ions with equal valences is determined by the ion-exchange constant ($K > 1$). The ion-exchange constant for
the case just considered is $K_K^{Na} = 0.333$; this predetermines the appearance
of sharp boundaries only for the exchange of sodium by potassium ions. For
rapid flow rates the criteria for the formation of sharp boundaries not
only depend on the ion-exchange constant, but also on the kinetics of the
process and geometric characteristics of the system. We shall consider
these aspects in more detail in the sections on dynamics. The cationite
exchange of an antibiotic with acidic properties - novobiocin (Figure 3.13)
- is an easily realizable substitution process.

Solutions of physiologically active substances are mostly neutralized
without difficulty. Cationites are used for organic acids and anionites
for organic bases, in accordance with the equations

$$RSO_3H^+ + K^+ + OH^- \rightleftharpoons RSO_3K^+ + H_2O, \qquad (3.5)$$

$$RCOOH + K^+ + OH^- \rightleftharpoons RCOO^-K^+ + H_2O, \qquad (3.6)$$

$$\overset{+}{RNOH^-} + H^+ + A^- \rightleftharpoons \overset{+}{RNA^-} + H_2O, \qquad (3.7)$$

$$RNH_2 + H^+ + A^- \rightleftharpoons \overset{+}{RNH_3}A^-. \qquad (3.8)$$

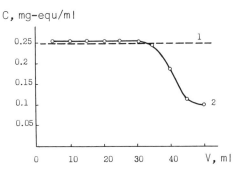

Fig. 3.12. Transformation of the sodium salt of benzylpenicillin to the potassium salt on permutite. Curves 1) concentration of benzylpenicillin leaving the column; 2) concentration of sodium leaving the column. Height of permutite layer in column 7 cm; solution flow rate 200 ml/h·cm².

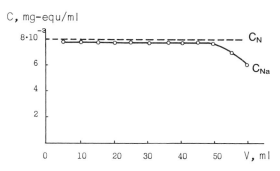

Fig. 3.13. Transformation of acidic novobiocin to the sodium salt on KB-4 ionite. C_N is the concentration of novobiocin leaving the column. C_{Na} is the concentration of sodium leaving the column. Column diameter 0.6 cm; ionite layer in column 8 cm high; solution flow rate 200 ml/h·cm².

The non-ionic absorption of organic ions in these conditions, as happens in the case of penicillin and synthetic organic cationites, is observed rarely. The feature of the processes (3.6) and (3.8), which occur in weak ionites, is that acidic solutions are not converted into ones with high pH's, or conversely going from high pH's to very low pH's is avoided in the presence of neutral salts. By contrast, ion exchange on sulfocationites, after having been neutralized with alkali, can be carried out with salts that result in the replacement of the cations by hydrogen ions without neutralizing them.

Practically useful neutralization is implemented under conditions that promote sharp interzone boundaries. It has been shown experimentally[124] that acid solutions are neutralized with large absorption capacity on anionites (in the basic form) that contain both weak and strong base groupings. The presence of many weak-base groupings, up to 7–8 mg-equ/g, a figure impossible for strong anionites, makes high sorption capacity possible, while the inclusion of strong-base groupings aids the formation of sharp interzone boundaries.

Ionite demineralization of solutions containing physiologically active substances includes two successive (or simultaneous) processes, viz., the exchange of metal ions for hydrogen ions on a cationite and the sorption of the anions onto an anionite while the solution is being neutralized following the scheme in (3.5) and (3.8), i.e.,

$$RSO_3^-H^+ + Me^+A^- \rightleftarrows RSO_3^-Me^+ + H^+ + A^-,$$

$$RNH_2 + H^+ + A^- \rightleftarrows R\overset{+}{N}H_3A^-.$$

Such a process is accomplished, like the demineralization of a streptomycin solution[132], on densely crosslinked cationites, such as SBS-1 or KU-2 × 20, and with small losses of antibiotic. As we have mentioned, it is convenient at the second stage to use a large-capacity polyfunctional anionite. In order to demineralize solutions of organic substances with acidic properties, densely crosslinked anionites must be used. Mixed filters containing both cationites and anionites at the same time may turn out, after the process has been finished, to be more convenient, because the solution's pH never changes. However, if the cation and anion exchange occur one after the other, the region in which the pH is changed only embraces the leading boundary, and this will be quite narrow if the sharp boundary criteria from the dynamic theory of frontal processes (Chapters 5 and 6) are satisfied.

Ionite demineralization is applied particularly when working with protein (enzyme) solutions and when removing low-molecular-weight components from biological liquors containing the cells and former elements of blood. In the latter case there is an additional requirement on the ionite, i.e., that damage to cell or subcell formations is avoided or reduced (e.g., hemosorbents). What lies behind the removal of small ions from a solution, viz., the ionite sieve method, can be used for the selective absorption of small organic ions when the ionite is impermeable to larger ions. This may reduce the toxicity of a solution, remove components that cause hypotension, or remove pyrogenetic or other admixtures.

3.4. ISOTHERMS OF EQUIVALENT ION EXCHANGE

When studying the separation or purification of substances or the fractionation of mixtures, concentration ratios are a measure of the efficiency of the process. High eluate concentrations are an important requirement for the sorption and desorption stages for preparations and production. Thus equations relating the concentrations in the solution and in the ionite at constant temperatures and pressures – ion-exchange isotherm equations – are needed. But first comes the question of whether it is valid to use the concept of concentration for the ionite phase. An ionite, and we are mainly considering synthetic network organic polyelectrolytes, is in fact a heterogeneous phase. Moreover, the heterogeneity is also predicted theoretically whenever there is copolymerization[19]. A strict analysis for these systems would be a statistical one in which an infinite set of states with different levels of interaction with the environment is considered for every component in the ionite. However, the concept of concentration of each of the components participating in the exchange and existing in the ionite phase is widely accepted for scientific investigation. It has been suggested that the repeated entities in the ion should be regarded not as individual sorbed ions states, but as whole sections of ionite that are large enough to contain all probable absorbed states, but small enough so that the whole system – set of entities – can be considered statistically homogeneous and to which the concept of concentration could be applied.

Ion-exchange isotherms for a process with two counter-ions may be obtained using electrochemical potentials. As the ions transfer there is, in addition to the work of expansion, an electric work due to the transfer of charge as each component is displaced. The heterogeneous equilibrium is given by equating the electrochemical (though not the chemical) potentials, i.e.,

$$\tilde{\mu}_i' = \tilde{\mu}_i'', \tag{3.9}$$

where

$$\tilde{\mu}_i = \mu_i^\circ(T,P) + RT \ln a_i + z_i \phi F;$$

$\mu_i^\circ(T,P)$ is the standard chemical potential,
a_i and z_i are the activity and charge of a component,
ϕ is the potential of a phase, and
F is Faraday's number.

One way of accounting for the third component, the transfer of the solvent, is to introduce an additional pressure, a pressure of swelling due to the limited swelling of the ionite. We shall consider the solvation effect later, but for some simple systems it is insignificant. In this case Equation (3.9) may, for double ionic exchange, be given as

$$\bar{\mu}_1^\circ + RT \ln \bar{a}_1 + z_1 \bar{\phi} F = \mu_1^\circ + RT \ln a_1 + z_1 \phi F$$
$$\bar{\mu}_2^\circ + RT \ln \bar{a}_2 + z_2 \bar{\phi} F = \mu_2^\circ + RT \ln a_2 + z_2 \phi F \tag{3.10}$$

or

$$RT \ln \frac{\bar{a}_1^{-1/z_1} a_2^{1/z_2}}{\bar{a}_2^{-1/z_2} a_1^{1/z_1}} = \frac{1}{z_1} \mu_1^\circ + \frac{1}{z_2} \bar{\mu}_2^\circ - \frac{1}{z_1} \bar{\mu}_1^\circ - \frac{1}{z_2} \mu_2^\circ, \tag{3.11}$$

which leads to the well-known form of the ion-exchange isotherm, i.e.,

$$\frac{\bar{a}_1^{-1/z_1} a_2^{1/z_2}}{\bar{a}_2^{-1/z_2} a_1^{1/z_1}} = e^{-\Delta\phi^\circ/RT} , \tag{3.12}$$

where $\Delta\phi^\circ$ is the standard potential for ion exchange. Note, however, that the analysis is for a two-component system in which each ion type is similarly hydrated in both coexistent phases. The transfer of the solvent in such an analysis may be accounted for in a model in which there is a variable (raised) pressure in the ionite phase. This pressure is due to the network structure hindering the transfer of ions within the ion-ite[136,146,160]. However, this method is rarely used nowadays for a thermodynamic analysis of ion exchange.

The above analysis is limited in that it is inconvenient to consider an ion (counter-ion), either solvated or unsolvated in the ionite, as a component. Firstly, a counter-ion can interact with fixed ionogenic groups in different ways. It is usually difficult to distinguish the fully ionized state of the fixed-ion/counter-ion system from an ion pair state, or even from the case of a weakly ionized state. This is particularly true of organic ions, which are weakly dissociated even in solution with other organic ions. Moreover, as we shall show later, organic ions interact with an ionite in an inherently complicated way, including a set of weak inter-actions, frequently hydrophobic, besides the ion-ion interactions. It is scarcely useful to divide these complexes into components. The idea of resinates is a widely used way of dealing with this problem. A resinate is a component in the ionite that includes the counter-ion and its immediate environment, both solvent and neighboring ionite. In the simplest case of ion exchange with two ions, the solvated exchanging ions in the solution and the corresponding resinates are considered to be the components. Note that another component (water) may be transferred during ion exchange. However, the whole water/counter-ion component transfer must be described by an additional coupled equation. For counter-ions this is the well-known equivalence equation. At the same time, it is difficult to introduce a coupling equation that establishes the stoichiometric relationship between the ions and the water. We suggest that in this respect it is easier in many systems to consider ion exchange to be a process that only involves the ions in the solution and the resinates, and include the transferred water as if it were part of the components. If the solvation level of each ion during the transfer from one phase to the other is independent of the ionic composition of the ionite, then this idea is justified. The resinate idea can be used to consider the solvated ions and the resinates as the components even when the ions are variably hydrated. Solvation itself can be considered independently of its role in ion exchange.

A thermodynamic analysis of equilibrium in heterogeneous systems is based on the superposition of thermodynamic functions for a closed system and functions of open subsystems that are part of the closed system. In the case of ion exchange the closed system includes both solution and ionite, while the subsystems are the ionite and solution separately. The following is valid for the closed system:

$$d\phi_{T,P} = d\bar{\phi} + d\phi = 0. \qquad (3.13)$$

For the open systems we have

$$d\bar{\phi} = \Sigma_i \bar{\mu}_i d\bar{n}_i - d\bar{E}',$$

$$d\phi = \Sigma_i \mu_i dn_i - dE', \qquad (3.14)$$

where $d\bar{E}'$ and dE' are the work of the electric forces during ion transport. If the independent transfers of each ion from one phase to the other are considered to be fluctuations, then the electric work must be considered for each such event, and this results in the equating of the electro-chemical potentials (Equation (3.9)). In this case the transferable component must be identical in material ratio in both phases; i.e., in the description above (Equations (3.9) and (3.12)) the ions in both solution and ionite must be in the same hydration state. This means the model may diverge from reality if the hydration state of the ionite changes during ion exchange.

Another possibility arises when ion exchange is considered using resinates. This analysis naturally precludes equating the chemical potentials, as this violates the condition for a closed system (3.13). Note that when the resinate model is used, the need for a standard component type in the solution and ionite loses its relevance. Furthermore, given equivalent ion exchange, when fluctuations correspond to the shifts of the stoichiometric ion exchange, the work of the electric forces can be avoided because in this process it is compensated by forward and backward ion transfers between the phases. For small displacements from equilibrium the following coupling equations can be introduced:

$$z_1 dn_1 = -z_2 dn_2, \quad z_1 d\bar{n}_1 = -z_2 d\bar{n}_2,$$

$$dn_1 = -d\bar{n}_1, \quad dn_2 = -d\bar{n}_2. \qquad (3.15)$$

For a four-component - two resinates and two ions in solution - ion exchange (a binary process in terms of the nominal ions) Equations (3.13) and (3.14) yield

$$\bar{\mu}_1 d\bar{n}_1 - \mu_2 dn_2 + \mu_1 dn_1 - \bar{\mu}_2 dn_2 = 0, \tag{3.16}$$

or using the coupling equation (3.15) give

$$1/z_1\bar{\mu}_1 - 1/z_2\bar{\mu}_2 + 1/z_2\mu_2 - 1/z_1\mu_1 = 0. \tag{3.17}$$

By using the relation between the chemical potential and activity of a component, viz., $\bar{\mu}_i = \bar{\mu}_i^\circ + RT \ln a_i$, $\mu_i = \mu_i^\circ + RT \ln a_i$, and proceeding from Equation (3.16) we obtain an equation that was first suggested and justified by Nikol'skii[167]. It is the most widely known equation for ion exchange and is analogous to Equation (3.12), i.e.,

$$\frac{\bar{a}_1^{-1/z_1} a_2^{1/z_2}}{\bar{a}_2^{-1/z_2} a_1^{1/z_1}} = K = e^{-\Delta\phi^\circ/RT}. \tag{3.18}$$

The method used here to obtain the equation shows that it can be used without limitation for any binary ion exchange. The resinates and solvated ions here are materially different components and the problem of liquid transport is avoided.

It is pertinent here to appraise the problems of "componentness," solvent transfer, and the effect of swelling. Firstly, in a binary ion exchange there are three independent components in each phase (ionite and solution), i.e., two electrolyte components and water (solvent) in the solution, and two resinates and water on the ionite. It is in principle possible to consider the ionite matrix as a component in the ionite phase. Conformational alterations of the network's structural elements do in fact contribute to a change in the thermodynamic potential during ion exchange. However, the use of resinates as components makes it possible to eliminate from consideration the transfer of water (solvent) and the energetic and entropic changes of the conformational alteration of the ionite matrix. The use of solvated states means that only two components, viz., the solvated ions, need be considered in the solution too. It should be remembered that by considering the closed system as a set of open systems (Equations (3.13) and (3.14)) presupposes the inclusion in Equation (3.16), on the basis of a well-defined coupling Equation (3.15), of only those components that necessarily participate in the interphase exchange. There have been attempts to include the transfer of water in analyses of ion-exchange isotherms either in the form of a thermodynamic swelling potential ($\Delta\phi_{sw}$) or one calculated from the chemical potential of water ($\bar{\mu}_w d\bar{n}_w + \mu_w dn_w$) [8,168].

We shall not use this approach here because an analysis of ion-exchange isotherms assuming the independent interphase transfer of water requires the consideration of the transfer of unsolvated counter-ions, and this creates certain difficulties when the models are used for calculations. Naturally, distinguishing between water brought by solvated ions and independently transferring water that solvates the ionite matrix is thermodynamically impermissible. Finally, with regard to "componentness," the use of four components for a binary ion exchange, i.e., the resinates and solvated counter-ions in the solution, when deriving and justifying the isotherm equations does not in any way exclude independent analyses of the three-component subsystems for the ionite and solution. As regards solvation and swelling, they are exceedingly significant. However, they must be considered by comparing (as we shall do later) a variety of systems, for example, ionites with different crosslink ratios. The conformational variability of network structures during ion exchange, for example, during ionite titration, has already been considered in Chapter 3 as an independent phenomenon. Returning to the ion-exchange isotherm Equation (3.18), we may note its universality to binary ion exchange for a given ionite-solution system.

When developing ion-exchange methods for preparative isolation and separation, the isotherm equation is used to estimate the sorption selectivity and to analyze the dynamics of column processes. This is done mainly to ascertain the conditions for the formation of sharp interzone boundaries using equilibrium dynamics and accounting for the kinetics using non-equilibrium dynamics theory. The conditions for complete column saturation by the substance being separated, for complete desorption, and for obtaining concentrated eluate can then be predicted. The data needed to deal with these topics do not have to be too accurate. Thus when calculating ion-exchange sorption, it is more than sufficient to obtain final results within 10%, especially bearing in mind the practical problems of creating systems and using process conditions under which the sorption selectivity constant approaches 100-1000 for organic ions (in most traditional processes this parameter rarely exceeds several dozen).

There are certain limits to the use of isotherms for studying the dynamics of ion exchange. Preparative ion-exchange chromatography (the dynamic processes of specific sorption and desorption) is based on the use of curved isotherms (isotherms of ion-exchange dynamics in the case of ion-exchange chromatography). A central topic here is thus the analysis of the form of the curvature, i.e., the sign of the second derivative of sorption capacity with respect to the substance's concentration in the external solution.

Some simplifications can be made to the ion-exchange isotherms, firstly for the solution phase. The isotherm equation (3.18) can be written as

$$\frac{\bar{C}_1^{1/z_1} C_2^{1/z_2}}{\bar{C}_2^{1/z_2} C_1^{1/z_1}} \cdot \frac{\gamma_1^{-1/z_1} \bar{\gamma}_2^{1/z_2}}{\gamma_2^{-1/z_2} \bar{\gamma}_1^{1/z_1}} = K, \qquad (3.19)$$

where C_1 and C_2 are the ion concentrations in the solution,
$\bar{C}_1$ and $\bar{C}_2$ are the ion concentrations in the ionite,
γ_1 and γ_2 are the activity coefficients of the ions in the solution,
$\bar{\gamma}_1$ and $\bar{\gamma}_2$ are the activity coefficients of the resinates in the ionite, and
z_1 and z_2 are the charges of the ions.

For ideal solutions the activity coefficients γ_1 and γ_2 are unity. For dilute electrolytes they may be assumed to be constant in solutions of constant ionic strength, right up to ionic strengths of 0.1, given the permissible error here (up to 10%). Under these conditions Equation (3.19) takes the form

$$\frac{\bar{C}_1^{1/z_1} C_2^{1/z_2}}{\bar{C}_2^{1/z_2} C_1^{1/z_1}} \cdot \frac{\bar{\gamma}_1^{-1/z_1}}{\bar{\gamma}_2^{-1/z_2}} = K_a, \qquad (3.20)$$

where K_a is the apparent ion-exchange constant.

Clearly K_a is a variable, and once the standard states are chosen for all the components it must still depend on the ionic strength of the solution.

The ionite (variable composition) phase is to some degree like a concentrated solution. Thus it is not possible to assume the activity coefficients are unity. However, the phase has some peculiarities quite different from those of a solution. The positions of the counter-ions and the resinates as a whole are fixed in space with restricted mobility. For some ionites this means that we can think of the influence of the micro-environment on the counter-ion and resinate states as being constant and independent of neighboring counter-ions and their resinates. Thus for the rigid, low-capacity aluminosilicates it has been suggested (and for some ion systems proved) that the counter-ion (resinate) activity coefficients in the ionite phase are constant. This is because each counter-ion inter-acts with a limited local constant environment that does not spread over to neighboring counter-ions and fixed ions, and hence a change in a neigh-boring counter-ion has no effect on the interaction energy between the counter-ion and its environment. The activity coefficients for these ionites can be taken as constant (both for the resinates and counter-ions) if the counter-ions are not large (for monovalent metal ions). The ac-tivities of resinates have been observed (experimentally) to be constant for the usual standard synthetic polymer sulfocationites with the simplest monoaminocarboxylic acids as the counter-ions. Quite reliable linear concentration relationships have been demonstrated for some amino acids, given that the activity coefficients are constant[189,190]. For these ionites and counter-ions using dilute solutions (γ_1 and γ_2 are constants) it is possible to assume the activity coefficients γ_1 and γ_2 are also constant. Equation (1.8) under these conditions takes the form

$$\frac{\bar{c}_1^{1/z_1} c_2^{1/z_2}}{\bar{c}_2^{1/z_2} c_1^{1/z_1}} = K_s, \tag{3.21}$$

where K_s is the selectivity coefficient.

The selectivity coefficient (also called the ion-exchange concen-tration coefficient) is constant, given changes in counter-ion composition in the ionite of a few percent, for only an insignificant number of systems. However, in order to tackle the practical problems indicated above it would be desirable to use Equation (3.21) when K_s is changed by less than 10-20%. This could be done for more systems if the analysis is limited to certain intervals of change in counter-ion composition in the ionite.

3.5. ION EXCHANGE ON IONITES WITH HETEROGENEOUS ACTIVE CENTERS

Synthetic network copolymers, a class that includes ionites, are by nature heterogeneous. Synthesis methods developed to produce isoporous ionites only reduce the heterogeneity. Hence it is not rigorous to con-sider an ionite as a homogeneous phase with a variable composition. Experimental data show that for dilute external solutions the concentration coefficient is significantly dependent on the ratio of the exchanging ions in the sorbed state. This indicates that there is an energetic (and entropic) heterogeneity of active centers. This heterogeneity depends on the heterogeneous distribution of the crosslinks in the copolymer[191], on the density heterogeneity within the ionite[192], and on the heterogeneous distribution of ionogenic groups that arises when they are incorporated into the copolymer[146,193-195], e.g., upon sulfonating, chloromethylating, or aminating. One way of analyzing the heterogeneity of ionites is to introduce the idea that each sort of counter-ion exists in the ionite in two states[191,196,197]. The concept of dissociated resinates[198], i.e., that counter-ions may exist, under the same conditions, either in a bound state with the fixed ions (e.g., as ion pairs) or in a dissociated state, belongs to this class of thermodynamic analysis. Note that a continuous series of states for a given ion[197] requires a large number of constants or extra relationships between the constants and that significantly reduces the usefulness of the ideas and exchanging the activity coefficients of the counter-ions in the ionite or resinate for dissociation constants. Adding even one state requires an additional dissociation constant, and even so activity coefficients may still be needed. It must be said that the simultaneous use of activity coefficients and the idea of dissociation is not in most cases useful because the former already can describe every aspect of the non-ideal behavior of an ionite (the varying of the concentration constants).

However, there are some ionites and ion-exchange systems for which it is useful to describe the equilibrium in terms of two resinate states. This is particularly true of weak ionites, the carboxylic acid cationites for example. In the last two sections we considered how the exchange capacity of a carboxylic cationite depended on the pH of the external solution. The variation meant that only some of the ionogenic groups can be occupied by the counter-ions in the solution. This is corroborated by dynamic (column) experiments. If a solution of electrolyte with a given pH is passed through a carboxylic cationite (hydrogen form), only some of the hydrogen atoms from the carboxylic groups are substituted for the incoming cations, which agrees with the data obtained by determining the exchange capacity by potentiometric titration in a solution with the same ionic strength and the same counter-ion. An ion exchange on a cation exchanger with two counter-ions whose bases yield the same shaped titration curves can be considered to be an equivalent ion-exchange on ionized ionogenic groups, the concentration of which is given by the exchange capacity and pH, and depends on the ionic strength of the solution.

A three-component ion-exchange on a carboxylic cationite can be analyzed using resinates. The ion-exchange equation for the system containing the hydrogen and one of the counter-ions, their resinate being dissociated, can be written as

$$\frac{\bar{C}_1^{1/z_1} a_H}{\bar{C}_H a_1^{1/z_1}} \frac{\bar{\gamma}_1^{-1/z_1}}{\bar{\gamma}_H} = K_{1,H}, \tag{3.22}$$

where $\bar{C}_1$ and $\bar{C}_H$ are the concentrations of the resinate containing the 1-counter-ion and the hydrogen, respectively. In the more traditional form for carboxylic cationites the equation is

$$pH = pK_1 + \log \frac{1 - \bar{\alpha}_1}{(z_1\bar{\alpha}_1)^{1/z_1}} - \frac{1}{z_1} \log a_1 + \log \frac{\bar{\gamma}_1^{-1/z_1}}{\bar{\gamma}_H}, \tag{3.23}$$

or for $z_1 = 1$:

$$pH = pK_1 + \log \frac{1 - \bar{\alpha}_1}{\bar{\alpha}_1} - \log a_1 + \log \frac{\bar{\gamma}_1}{\bar{\gamma}_H}, \tag{3.24}$$

where $\bar{\alpha}_1$ is the dissociation level of the cationite or the quantity of the 1-cation that has been sorbed given an activity a_1 of the counter-ion in the solution.

By analogy we get the following for the hydrogen/2-counter-ion system (this also forms a dissociated resinate):

$$pH = pK_2 + \log \frac{1 - \bar{\alpha}_2}{(z_2\bar{\alpha}_2)^{1/z_2}} - \frac{1}{z_2} \log a_2 + \log \frac{\bar{\gamma}_2^{-1/z_2}}{\bar{\gamma}_H}, \qquad (3.25)$$

or for $z_2 = 1$:

$$pH = pK_2 + \log \frac{1 - \bar{\alpha}_2}{\bar{\alpha}_2} - \log a_2 + \log \frac{\bar{\gamma}_2}{\bar{\gamma}_H}. \qquad (3.26)$$

If we consider the triplet system that includes the three cations in solution ($H-cat_1-cat_2$) and the corresponding resinates, then we find from the general principles concerning the interactions of the three doublet systems[75] that we get the following for the cations cat_1 and cat_2:

$$\frac{\bar{a}_1^{-1/z_1} a_2^{1/z_2}}{\bar{a}_2^{-1/z_2} a_1^{1/z_1}} = K_{1,2}. \qquad (3.27)$$

It is simpler to obtain Equation (3.27) for ion exchange of the H-Na form of a carboxylic cationite and for the sorption of organic ion (Na-organic ion exchange) when the ionization of the ionite is constant.

Thus the real task of studying the exchange of two cations in strong bases and on carboxylic cationites can be considered to be the analysis of two counter-ions according to Equation (3.27). The feature of this three-component system is that it can be reduced to the desirable two-component one with a minimum allowance being made for the third component, viz., hydrogen. We shall do this for a constant pH. However, as we can see from Equations (3.25) and (3.26), the ionization, i.e., the quantity, of the sorbed ions in the H-cat_1 and H-cat_2 two-component systems are dependent on the concentration of the cations in the external solution and the activity coefficients of the resinates (their interactions), at least in the micro-environment. Thus the exchange capacity, for the cat_1-cat_2 ion exchange, is in principle a variable quantity. However, as we shall demonstrate later, the background of a persistent exchange capacity (persistent number of exchanged groups in the carboxylic cationite occupied by hydrogen) can be taken as being relatively constant for a given pH.

It is often necessary to introduce the concept of relative exchange capacity (Section 4.2) for the ion exchange of complex organic ions. The ionite is thus considered as a heterogeneous phase in which only some of the ionogenic groups can be reached by the organic ions[8,11]. Any equilibrium (hence thermodynamic) analysis must therefore account only for the concentration of resinate participating in the ion exchange. Thus the ion-exchange equation (3.19) must be written as

$$\frac{\bar{C}_1^{-1/z_1} \bar{C}_2^{-1/z_2}}{(M - \bar{C}_1)^{1/z_2} C_1^{1/z_1}} \frac{\bar{\gamma}_1^{-1/z_1} \gamma_2^{1/z_2}}{\bar{\gamma}_2^{-1/z_2} \gamma_1^{1/z_1}} = K, \qquad (3.28)$$

where M is the concentration of resinate that can be reached by the large (organic) ions.

For systems that include organic ions it is more convenient to use the more accepted notation of

$$\bar{C}_1 = m_1, \quad M - \bar{C}_1 = M - m_1 = m_2.$$

To use Equation (3.28) M must be determined for a static or dynamic experiment, though in doing so attention must be paid to the usual gradual slowing in the rate of the inter-phase exchange that occurs in ionites with limited permeability for large ions. Tests show that the diffusivities in sorbent beads fall gradually from 10^{-7}–10^{-8} cm^2/s to 10^{-10}–10^{-11} cm^2/s. To complete a process in these systems with beads 200–300 μm in radius requires 5–20 h. The error in the determination of $\bar{C}$ is a few percent, and this does not distort a theoretical analysis of a preparative process, and even less so for an industrial one. It must be taken into account that the purity of the antibiotic, alkaloid, hormone, or enzyme ions used for experimental studies is 95–97%, as a rule, though often it is even less, especially for proteins (even crystals). Naturally, we need to know here how to assess the equilibrium, which may be either a true or metastable one. Remember above all that the thermodynamic analysis for substances and systems in a "frozen" state is legitimate and widely used for amorphous (particularly glassy) states and for a host of organic substances for which the activation energy for the alteration is an order of magnitude less than that of thermal motion. Once the apparent sorption has ceased (5–20 h) for the ion exchange of organic ions, and a relative absorption capacity of 50–80% (e.g., during antibiotic sorption) has been achieved, no further loading of the inaccessible functional groups is observed, even after hundreds of hours. The accessibility of the ionogenic groups, as regards the organic ions, can be altered by grinding the ionite, or by changing the temperature, solution pH, or ionic strength[15,199]. In the final analysis for real systems that are of practical importance the ionite cannot be ground to make the functional groups accessible to all counter-ions. Moreover, mechanical grinding is a treatment of a polyelectrolyte that leads, as we shall show, to many considerable changes in its thermodynamic properties and in the ion-exchange processes that take place in the ionite. Thus it becomes necessary to study ion exchange given relative exchange capacities of less than 100% if it is impracticable to get an ionite all of whose ionogenic groups are accessible. Obviously, the relative exchange capacity should be kept reliably within acceptable levels, and ionites of a given size should be investigated for constant (or nearly constant) temperatures, and constant ionic strengths and solution pH's.

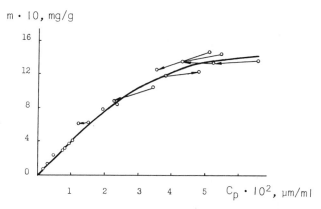

Fig. 3.14. Sorption isotherm for vitamin B$_{12}$ on Biocarb-T starting from different ratios of initial to equilibrium concentrations.

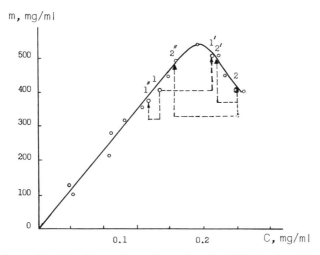

Fig. 3.15. Sorption isotherm for adenosine by SNK ionite for different
initial to equilibrium concentration ratios.

It is desirable for these systems to be tested for process revers-
ability given changes in the counter-ion concentration in the external
solution by approaching the equilibrium starting from different concentrations. The process of approaching the sorption equilibrium on a carboxylic cationite is shown in Figures 3.14 and 3.15 for vitamin B_{12} and
adenosine. The initial concentrations of the organic ions in the external
solution were both above and below the equilibrium concentrations. As can
be seen, the systems tend to the same equilibrium irrespective of the
initial concentration ratios. This is even true when the isotherm is very
complicated with a maximum, which is taken to be due to association in the
external solution of the sorbates.

3.6. THE SELECTIVITY OF ION EXCHANGE AS A FUNCTION OF THE COUNTER-ION
MOLE FRACTION IN THE IONITE

Real ion-exchange, being heterogeneous with respect to the ionite
phase, must be studied taking into account the resinate activity coefficients. However, it is sometimes possible to create and study an ionite
that for one sort of counter-ion has the properties of an ideal phase,
i.e., constant activity coefficients. In other words, the selectivity of
sorption is constant for changes in the mole fraction of the counter-ion in
the ionite when it is in contact with a dilute electrolyte solution. This
class of ionites contains those with small exchange capacities and widely
separated ionogenic groups, and some ionites whose counter-ions are the
simplest amino acids[200].

Considerable divergences from ideality is more often the rule, this
being particularly true when organic ions are being sorbed that interact
with each other because of their large sizes and thus get entangled in
neighboring resinates. Curve 1 in Figure 3.16 is the curve for an ideal
system, while curves 2 and 3 are for systems in which the selectivity
either fell or rose when counter-ion 1 was introduced. Curve 2 is typical
for most systems with small counter-ions (e.g., metal ions); there is a
fall in K as the mole fraction, in the ionite, of the more selectively
sorbed 1-ion increases. This corresponds to the fixed ions being heterogeneous in energy and to the ionite's ionogenic groups being filled
preferentially so that the group filled first causes the thermodynamic
potential to fall the most.

At each subsequent stage the selectively sorbed ion interacts with
the ionite less selectively. The opposite effect is observed experimentally for a considerable number of organic ions. Here the sorption
selectivity grows in proportion to the number of organic counter-ions in
the ionite (curve 3, Figure 3.16), and a high sorption selectivity is
typical in competition with the 2-ions. This effect should be classed
with the phenomenon of cooperativity which appears in these ion-exchange
systems due to changes in the micro-environment for the organic counter-
ions and is due to "additional" interactions between pairs of organic
counter-ions. The appearance of an organic counter-ion in a neighboring
position can raise the energy (and often the system's entropy, as we
shall show in Section 4.10) with which the counter-ion interacts with its
immediate environment, which includes both fixed and counter ions. This
leads to the increase in the selectivity of the ion exchange. Thus we
are justified in calling the counter-ion/ionite interaction corresponding
to curve 3 cooperative instead of statistical, for which curve 2 is characteristic. Experimental confirmation of this can be seen in Figure 3.17.
Both the cooperative and the statistical effects can be observed here for
the sorption of oxytetracycline onto sulfocationites with different exchange capacities[117]. For the sorption of tetracycline onto ionites
with large exchange capacities, i.e., ones with closely distributed
ionogenic groups, the sorption selectivity grows in proportion to the
loading of the ionite with tetracycline; i.e., cooperative interaction is
observed. The situation is considerably different after the same ionite
is partially desulfonated. The cooperative interactions disappear, and all
the properties of a statistical system with energetically heterogeneous
ionogenic groups appear. The sorption selectivity for the oxytetracycline
falls in proportion to the quantity of the organic ion in the ionite.
Even more complicated relationships between the selectivity of organic ions
and their mole fractions in the ionite are known. Thus for oleandomycin
(Figure 3.18), curves with a minimum that should be interpreted as the
appearance of the cooperative effect only arise after there is a large
quantity of the organic counter-ion sorbed.

The energetic and entropic heterogeneity of an ionite may be analyzed
by studying how activity coefficients of the resinates depend on the loading of the ionite with a given counter-ion. In our thermodynamic analysis
of ion exchange using resinates, we need not consider the solvent (water)
as an independent component for the exchange of two ions. It is suffi-
ciently rigorous to account for only two types of resinate and the two

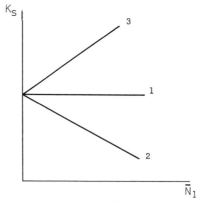

Fig. 3.16. Ion-exchange sorption selectivity versus mole fraction of
organic counter-ion in ionite. Curves 1) ideal isotherm;
2) statistical isotherm; 3) cooperative isotherm.

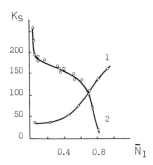

Fig. 3.17. Sorption selectivity of oxytetracycline versus mole fraction of antibiotic in ionite on Dowex-50 × 1 sulfocationites with different exchange capacities. Curves 1) 5.0 mg-equ/g; 2) 2.3 mg-equ/g.

exchanging ions in the solution. We can thus use a variation of Equation (3.19), i.e.,

$$\frac{\bar{N}_1^{1/z_1} c_2^{1/z_2}}{\bar{N}_2^{1/z_2} c_1^{1/z_1}} \; \frac{\bar{\gamma}_1^{-1/z_1} \gamma_2^{1/z_2}}{\bar{\gamma}_2^{-1/z_2} \gamma_1^{1/z_1}} = K, \tag{3.29}$$

where $\bar{N}_1$ and $\bar{N}_2$ are the mole fractions of the resinates in the ionite. Obviously the activity coefficients $\bar{\gamma}_1$ and $\bar{\gamma}_2$ in Equation (3.29) differ from those in Equation (3.19). However, we shall keep the same notation. We shall do the analysis for a system that has already been worked out [201] and give Equation (3.29) as

$$K = K_a \frac{\bar{\gamma}_1^{-1/z_1}}{\bar{\gamma}_2^{-1/z_2}} \tag{3.30}$$

or

$$\ln K = \ln K_a + \frac{1}{z_1} \ln \bar{f}_1 - \frac{1}{z_2} \ln \bar{f}_2 \tag{3.31}$$

or

$$-d \ln K_a = \frac{1}{z_1} d \ln \bar{f}_1 - \frac{1}{z_2} d \ln \bar{f}_2. \tag{3.32}$$

By combining Equation (3.32) with the Gibbs–Duhem equation for the ionite phase, i.e.,

$$\bar{N}_1 d \ln \bar{f}_1 + \bar{N}_2 d \ln \bar{f}_2 = 0, \tag{3.33}$$

we get

$$\frac{1}{z_1} d \ln \bar{f}_1 = \frac{(-1/z_1)\bar{N}_2}{(1/z_2)\bar{N}_1 + (1/z_1)\bar{N}_2} d \ln K_a \tag{3.34}$$

and

$$\frac{1}{z_2} d \ln \bar{f}_2 = \frac{(1/z_2)\bar{N}_1}{(1/z_2)\bar{N}_1 + (1/z_1)\bar{N}_2} d \ln K_a. \tag{3.35}$$

65

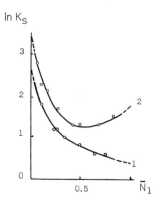

Fig. 3.18. Sorption selectivity of oleandomycin on Dowex-50 × 1 versus equilibrium ionic strength. Curves 1) 0.003 mg-equ/ml; 2) 0.06 mg-equ/ml.

In order to get the activity coefficients for Equations (3.34) and (3.35) we must choose standard states for each of the ions. Most often the standard state is chosen to be when the ionite is completely loaded with the given counter-ion, i.e.,

$$\ln \bar{f}_1 = 0, \quad \bar{N}_1 = 1 \ (\bar{N}_2 = 0), \quad x = 1,$$

$$\ln \bar{f}_2 = 0, \quad \bar{N}_2 = 1 \ (\bar{N}_1 = 0), \quad x = 0,$$

(3.36)

where x is the equivalent fraction of the second counter-ion in the ionite or of the second resinate:

$$x = \frac{(1/z_1)\bar{N}_2}{(1/z_2)\bar{N}_2 + (1/z_1)\bar{N}_2} .$$

(3.37)

Integrating (3.34) and (3.35) for these boundary conditions enables us to determine the activity coefficients of the resinates from the experimental curves for the K_a-x relationship, i.e.,

$$\ln \bar{f}_1 = z_1 \int_x^1 \ln K_a \ dx - z_1(1 - x) \ln K_a$$

(3.38)

or

$$\ln \bar{f}_2 = z_2 x \ln K_a - z \int_0^x \ln K_a \ dx.$$

(3.39)

For systems that include two equally charged but differing counter-ions Equations (3.38) and (3.39) take the form

$$\ln \bar{f}_1 = \int_{N_1}^1 \ln K_a \ d\bar{N}_1 - (1 - \bar{N}_1) \ln K_a$$

(3.40)

and

$$\ln \bar{f}_2 = \int_0^{N_2} \ln K_a \ d\bar{N}_2 + (1 - \bar{N}_2) \ln K_a.$$

(3.41)

The calculation of the thermodynamic constants of ion exchange is of great practical importance. Substituting the activity coefficients from Equations (3.38) and (3.39) into Equation (3.31) yields

$$\ln K = \int_0^1 \ln K_a \, dx, \qquad (3.42)$$

and for $z_1 = z_2$ we get

$$\ln K = \int_0^1 \ln K_a \, d\bar{N}_1. \qquad (3.43)$$

Other approaches to the calculation of the activity coefficients and ion-exchange constants have been worked out. The same ionite state when one of the counter-ions is absent is taken as the standard state for both the resinates[202]. The need to consider a three-component system, one including a solvent, loses its significance for a thermodynamic analysis of a given ionite and a given counter-ion pair when using the concept of resinate that we have introduced.

The ion-exchange constants are usually experimentally determined using weak ionic solutions. It is then possible to substitute K_a for K_s in Equations (3.42), (3.43), and (3.38)-(3.41). It is quite sufficient for most systems involving organic ions to estimate the ion-exchange coefficients, and this can be done by measuring the selectivity constants given a constant ionic strength of the external solution that is less than 0.1.

3.7. SORPTION SELECTIVITY OF IONS AS A FUNCTION OF ELECTROLYTE CONCENTRATION IN THE EXTERNAL SOLUTION

The sorption selectivity of organic ions, particularly complex ones like those of physiologically active substances, is often very much larger than the selectivity of the sorption of metal ions. Thus the selectivity of the sorption of tetracycline on some sulfocationites can be hundreds of times larger than that of the sorption of sodium or hydrogen, while the sorption selectivity of novobiocin can be one to two thousand times that of chloride.

The large selectivities mean that large quantities of antibiotics, alkaloids, hormones, enzymes, etc. can be sorbed from many extracts (directly from culture liquors or after they have been treated or filtered, these being called native solutions), even when there are a great many competing ions, particularly mineral salts. For example, after the biosynthesis of the tetracycline group of antibiotics, a solution is obtained with an ionic strength with respect to mineral ions of about 0.1-0.2 mg-equ/ml at an antibiotic concentration of 0.01-0.05 mg-equ/ml. Given the high ion-exchange constants, the small ions in the solution have a limited ability to compete in the ion-exchange sorption. As a result, antibiotics are sorbed from their active culture liquors mainly in com-petition with other organic counter-ions. These relationships give a series of highly specific sulfocationites the ability to sorb up to a gram of antibiotic per gram of sorbent directly from a native solution. This parameter value is also achieved for the sorption of other antibiotics and physiologically active substances. In particular high competition has been demonstrated for the sorption of enzymes[115]. It should be noted that the increase in the ionic strength due to the competing mineral ions may both affect the ion-exchange equilibrium as defined by the ion-exchange iso-therm, and lead to a reduction in the swelling of the ionite which may render some parts of the ionite inaccessible to the large ions.

Variations in the concentration of the organic ions in the external solution for a constant mineral ion concentration have been observed both in extracts from animal and vegetable sources, and in the culture liquors resulting from microbe synthesis. If these complicated mixtures are modelled by a binary ion-exchange with organic ions (C_1 and $\bar{C}_1$) and mineral ions (C_2 and $\bar{C}_2$), then the ion-exchange equation for the exchange of equally charged ions takes the form

$$\frac{\bar{C}_1}{M - C_1} = Kf(\gamma) \frac{C_1}{C_2},\tag{3.44}$$

where M is the limiting sorption capacity for the organic ions. If C_2 is the concentration of the small competing ions and is constant, then the ion-exchange isotherm will take the form

$$\bar{C}_1 = \frac{Kf(\gamma)MC_1}{C_2 + Kf(\gamma)C_1}.\tag{3.45}$$

Thus the $\bar{C}_1$-C dependence will take the form of an isotherm displaying saturation, i.e., a curve that rises steeply at first and then levels out at some limit.

For a native solution obtained from the culture liquor for tetracycline, the way the quantity of sorbed antibiotic depends on its activity (concentration) in the solution takes the form of an isotherm displaying saturation, given that the only variable is this concentration. At low tetracycline concentrations in the solution the sorption capacity grows linearly with concentration, though it increases more slowly later and tends to a constant value. This means that when working with fairly inactive culture liquors it is necessary to use the same number of columns (or the same quantity of sorbent) at a large scale for sorbing the antibiotic from a given volume of liquid. When the activity of the antibiotic in the solution grows, which it starts doing at a concentration of 2-3 mg/ml for tetracycline, then the quantity of sorbent used must rise in proportion to the antibiotic concentration.

Concentration ratios are exceedingly important for the sorption selectivity of an ion during the ion exchange of ions with differing charges. Consider a change in the ratios of the quantity of sorbed ions in a dilute solution given that the ratio between their concentrations is constant ($C_1/C_2 = P$). The ion-exchange isotherm may be written in this case as

$$\frac{\bar{C}_1^{1/z_1}}{\bar{C}_2^{1/z_2}} = K_s \frac{\gamma_1^{1/z_1}\bar{\gamma}_2^{1/z_2}}{\gamma_2^{1/z_2}\bar{\gamma}_1^{1/z_1}} P^{1/z_2} C_1^{(1/z_1 - 1/z_2)} = A\tag{3.46}$$

or

$$\frac{\bar{C}_1^{1/z_1}}{(M - \bar{C}_1)^{1/z_2}} = K_s \frac{\gamma_1^{1/z_1}\bar{\gamma}_2^{1/z_2}}{\gamma_2^{1/z_2}\bar{\gamma}_1^{1/z_1}} P^{1/z_1} C_1^{(1/z_1 - 1/z_2)} = A.\tag{3.47}$$

Given a dilute external solution, the quantity of sorbed ion can change. At large dilutions we have

$$\left.\begin{array}{l} \text{for } z_1 < z_2, \; m_{1,C_1\to0} = 0, \; m_{2,C_1\to0} = m, \\[2mm] \text{for } z_1 > z_2, \; m_{1,C_1\to0} = M, \; m_{2,C_1\to0} = m - M. \end{array}\right\}\tag{3.48}$$

The basic system we shall consider is an organic-mineral ion pair, though any pair of a large and small ion will do. The first ion (C_1, $\bar{C}_1$) is the organic (large) ion. The limiting sorption capacity for the first ion type is M, which is the total number of active centers accessible for this ion, while the sorption capacity for the second ion is the overall number of ionized centers in the ionite. According to (3.48), the relative quantity of the ions with the large charge increases as the solution is diluted. For ideal systems (γ_1, γ_2, $\bar{\gamma}_1$, $\bar{\gamma}_2$ are constants) Equations (3.46) and (3.47) offer the simplest way of calculating changes in m_1 and m_2 given dilute or concentrated solutions. For a substantially non-ideal ionite phase, dilution leads to an additional increase in the number of sorbed multi-charged organic ions ($z_1 > z_2$) or to an additional reduction in the number of organic ions with low charge ($z_1 < z_2$) given the cooperative effect between ions in the ionite (see Figure 3.16) or to the inverse influence of the non-ideality of the ionite phase given statistical ionite/counter-ion interactions on heterogeneous ionites. The data in Table 3.2 demonstrate that diluting solutions containing streptomycin hydrochloride and sodium chloride considerably increases the quantity of streptomycin sorbed on the carboxylic cationite CPA. However, this change is significantly lower than what would be expected for an ideal system. This must be primarily due to the energetic heterogeneity of the ionite.

The dilution or concentration of solutions containing ions with different valences is of great practical importance in sorption processes. For example, diluting the solution when working with culture liquors, native solutions, or extracts substantially increases the absorption capacity of the ionites for systems involving antibiotics such as streptomycin, neomycin, or kanamycin, which exist in the solution as multi-charged ions. This is because on average the charge on these ions is greater than those on competing ions. Although it may not be commercially practical to increase the volume of solution passing down a column on an industrial scale, dilution is a very promising technique for laboratory-scale antibiotic preparations, especially when the next step is a chemical analysis. The reverse problem crops up here, viz., extracting mineral ions from a physiologically active substance, i.e., demineralization[132]. The initial concentration of a solution containing multi-charged organic ions reduces the losses encountered when demineralizing it on a ionite (a molecular or ionite sieve), the beads of which are fairly inaccessible to large ions.

The use of low-capacity ionites of optimum homogeneity, and dilute solutions allows us to use Equations (3.46) and (3.47) to calculate the charge of one of the counter-ions, if the charge on the other is known. Given a rational choice of system, non-ideal effects have little effect on the calculation.

Table 3.2. Sorptive Capacity of KFU Carboxylic Cationite for Streptomycin in the Presence of Sodium Ions in Dilute Solutions

Streptomycin concentration (mg-equ/ml$\cdot 10^3$)	Sodium ion concentration (mg-equ/ml)	m_{strep} (mg-equ/g)
5.17	1.50	0.113
2.58	0.75	0.796
1.03	0.30	1.92
0.52	0.15	2.91

3.8. ION EXCHANGE WITH WEAK ELECTROLYTES

Our thermodynamic analysis of ion exchange (Section 3.4) was based on the concentration of the ions in the solution and the concentration of the resinates in the ionite. However, when studying the behavior of weak electrolytes it is better, sometimes essential, to work with the concentration of the electrolytes in the solution rather than ion concentrations.

We shall consider the exchange of two counter-ions, one of which (the 1-ion) forms a weak electrolyte together with co-ions, the electrolyte being partially dissociated in the external solution. Most often these systems include an organic and a mineral ion. The ion-exchange equation can now be written as

$$\frac{\bar{c}_1^{1/z_1} c_2^{1/z_2}}{\bar{c}_2^{1/z_2} (\alpha_1 c_1^*)^{1/z_1}} \frac{\bar{\gamma}_1^{-1/z_1} \gamma_2^{1/z_2}}{\bar{\gamma}_2^{-1/z_2} \gamma_1^{1/z_1}} = K, \qquad (3.49)$$

where c_1^* is the concentration of the weak electrolyte in the solution, and α_1 is its degree of ionization, or as

$$\frac{\bar{c}_1^{1/z_1}}{\bar{c}_2^{1/z_2}} \frac{c_2^{1/z_2}}{c_1^{*1/z_1}} = K \alpha_1^{1/z_1} f(\gamma) = K_s, \qquad (3.50)$$

where

$$f(\gamma) = \frac{\gamma_1^{1/z_1} \bar{\gamma}_2^{-1/z_2}}{\gamma_2^{1/z_2} \bar{\gamma}_1^{-1/z_1}} .$$

We shall consider the selectivity of the ion-exchange sorption (K_s) to be a function of the solution pH. The way K_s depends on the pH is determined by both α_1 and $f(\gamma)$. It is perfectly justifiable to assume that the influence of α_1 significantly exceeds that of $f(\gamma)$ in the vast majority of systems. The actual form of the α_1-pH relationship depends on the number of ions formed when the weak electrolyte dissociates, and on the ionic strength of the solution. For a dilute solution and an electrolyte with one degree of dissociation for the pH range in question, the relationship between α_1 and the concentration of hydrogen ions in the solution has the form

$$\alpha_1 = \frac{C_{H^+}}{K_{DB} + C_{H^+}} , \quad \alpha_1 = \frac{K_{DA}}{K_{DA} + C_{H^+}} , \qquad (3.51)$$

where K_{DB} is the dissociation constant for a base (cation exchange), and K_{DA} is the dissociation constant for an acid (anion exchange).

A comparison of Equations (3.50) and (3.51) enables us to predict that the sorption selectivity of the anions of a weak electrolyte (anion exchange) will grow, while that of the cations (cation exchange) will fall, as the solution pH is increased.

Quite a bit of attention is being paid to the effect of pH on the sorption of weak electrolytes onto weak ionites[15]. The most probable postulate for many ionites and counter-ions is that ion exchange only takes place on the ionized functional groups of the ionite, i.e., the hydrogen-type resinates for cationites and base-type resinates for the anionites do not participate in the ion sorption by ion exchange. Since both counter-

ions undergo ion exchange to some extent, and since our task is to explain
how the sorbed concentration of one of the ions depends on the acidity of
the solution, we must introduce the limitation that the concentration of
the second counter-ion (C_2) is constant. Thus Equations (3.49) and (3.50)
for the weak electrolyte, and Equation (3.45) for the weak ionite take the
form

$$\bar{C}_1 = \frac{K\alpha_1 f(\gamma)\bar{M}\bar{\alpha}C_1}{C_2 + K\alpha_1 f(\gamma)C_1} \, , \tag{3.52}$$

where M is the exchange capacity for the 1-ion given the ionization of all
of the ionite's functional groups, and
$\bar{\alpha}$ is the degree of ionization of the ionite's ionogenic groups.

The analysis of Equation (3.52) is hindered by the relationship between the
ionite ionization and the counter-ion type, which we considered   earlier
(see Section 3.1). This relationship, however, has a weaker influence than
the effect of the pH, whatever the counter-ion and given a weakly ionized
ionite.

Equation (3.52) can be analyzed for the influence of pH on $\bar{\alpha}_1$ either
for arbitrary values of C_1 and C_2 or for the limiting cases of

$$C_2 \ll K\alpha_1 f(\gamma)C_1 \quad (1),$$

$$C_2 \gg K\alpha_1 f(\gamma)C_1 \quad (2). \tag{3.53}$$

For small concentrations of the competing ions (C_2) or for large C_1 the
first case in (3.53) simplifies Equation (3.52) to

$$\bar{C}_1 = \bar{M}\bar{\alpha}. \tag{3.54}$$

This indicates that the concentration of the sorbed 1-ion is monotonically
dependent on the pH, or on the loading of the ionogenic groups by the
1-counter-ion. In cation exchange an increase in the solution pH raises
the concentration of the sorbed ion formed by the dissociation of the weak
electrolyte, while the reverse occurs for anion exchange. This limiting
case (the first of (3.53)) rarely occurs. Much more often the 1-ion is
sorbed with significant constant concentrations of the competing 2-ions.
Equation (3.52) then simplifies to

$$C_1 = \frac{K\alpha_1 f(\gamma)\bar{M}\bar{\alpha}C_1}{C_2}. \tag{3.55}$$

The curve corresponding to the $\bar{C}_1$-pH relationship has a maximum because a
change in α_1 or $\bar{\alpha}$ is antibatic both for anion and cation exchange, if the
change is considered to be a function of the solution acidity.

We shall now consider the ion exchange of the cations of a weak (1)
and a strong (2) electrolyte on a weak (e.g., carboxylic) cationite. By
using Equation (3.51) and the second case in (3.53) we can transform
Equation (3.55) into a form that gives a direct relationship between the
concentration of sorbed 1-ions and the acidity of the medium given the
constant concentrations C_1 and C_2, i.e.,

$$\bar{C}_1 = \frac{K[C_{H^+}/(K_{DB} + C_{H^+})]f(\gamma)M10^{[(pH - pK_{typ})/n]}C_1}{(1 + 10^{[(pH - pK_{typ})/n]})C_2}. \tag{3.56}$$

An analysis of Equation (3.56) for $pK_{DB} > pK_{typ}$ (a limitation introduced for weakly dissociated electrolytes in solution) shows that either a significant increase or a significant reduction in the solution acidity reduces the sorbability of the 1-ions, i.e.,

$$\text{for pH} \gg pK_{DB}, \quad \alpha_1 \to 0, \quad \bar{\alpha}_1 \to 1, \quad \bar{C}_1 \to 0,$$

$$\text{for pH} \ll pK_{DB}, \quad \alpha_1 \to 1, \quad \bar{\alpha}_1 \to 0, \quad \bar{C}_1 \to 0. \tag{3.57}$$

The experimental evidence for this maximum is provided in many papers[15]. The majority of the experimental papers do not consider the relationship between the concentrations C_1 and C_2, which our analysis indicates could affect the shape of the $\bar{C}_1$-pH curve and lead to the disappearance of the maximum.

3.9. ION EXCHANGE, HYDRATION, AND SWELLING

Organic ions, particularly those of physiologically active substances, are only sorbed in large quantities by very permeable ionites. The hydration of macroporous ionites has little effect in this respect. However, when organic ions are selectively sorbed onto gel or heteronet ionites, the ionite is usually compacted or dehydrated somewhat, and this has an effect on the energetics of the ion exchange. As we noted earlier, the thermodynamics of ion exchange with a given ionite and exchanging ions can be considered without an account of the extra component (water). However, the component must be accounted for when comparing different ionites, such as those with differing crosslink ratios.

The complete dehydration of gel and heteronet ionites by desiccation at room temperature (even more so when heated) can lead to irreversible changes in the ionite's net structure and reductions in its permeability. Thus in order to preserve the porosity of a network polyelectrolyte it should be dehydrated under vacuum and at low temperatures. Some gel ionites (and macroporous ones in particular) change their structure reversibly when air-dried at room temperature. The n_1-α hydration isotherms of heteronet and gel ionites mostly differ to the right of the curve (Figure 3.19), n_1 being the number of moles of hydrating water per milligram equivalent of ionite, and α being the relative vapor pressure of the water. The slower growth in n_1 with the growth in α in the 0.8-1.0 interval is typical of heteronet ionites. This is related to the small quantity of free water in the ionite.

The thermodynamic swelling potential of the ionite may be obtained from the chemical potentials of the ionite and water, given as components, i.e.,

$$\Phi_{sw} = n_1\mu_1 + n_2\mu_2 - n_1\mu_1^o - n_2\mu_2^o, \tag{3.58}$$

where μ_1 and μ_1^o are the chemical potentials of the water in the swelled ionite and in the free state, and
μ_2 and μ_2^o are the chemical potentials of the ionite before and after swelling.

We have for water

$$\mu_1 = \mu_1^o + RT \ln \alpha, \tag{3.59}$$

where α is the relative vapor pressure of water, i.e., the ratio of the vapor pressure of water in equilibrium with the swelled ionite to the vapor

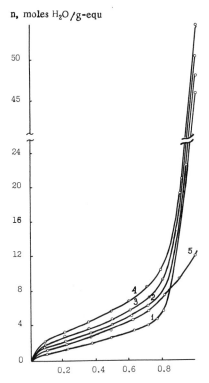

Fig. 3.19. Isotherm for sorption of steam by sulfocationites. Copolymers of N-methacryloyl-(4-aminobenzenesulfonic acid) and 1) 5 mol.%, 2) 9 mol.%, 3) 16 mol.%, and 4) 41 mol.% HMDMA; and 5) KU-2 (8% DVB) sulfocationite.

pressure of water in the standard state at the same temperature and pressure.

Using the Gibbs-Duhem equation

$$\Sigma n_i d\mu_i = O(T,P)$$

we get the following for a two-component system:

$$n_2(\mu_2 - \mu_2^0) = - \int_{-\infty}^{\mu_1} n_1 d\mu_i = -RT \int_0^\alpha \frac{n_i}{\alpha} \, d\alpha. \tag{3.60}$$

A comparison of (3.58), (3.59), and (3.60) yields an expression that can be used to obtain the swelling potential of an ionite from experimental data, i.e.,

$$\Phi_{sw} = n_1 RT \ln \alpha - RT \int_0^\alpha \frac{n_i}{\alpha} \, d\alpha. \tag{3.61}$$

A comparison of Φ_{sw} for ionites containing exchanging counter-ions in different ratios enables us to estimate the differential thermodynamic potential of swelling, i.e.,

$$\Delta\Phi_{sw} - \frac{\gamma\Phi_{sw}}{\gamma\bar{N}_1}, \tag{3.62}$$

where $\bar{N}_1$ is the mole fraction of the first (organic) counter-ion in the ionite.

73

Table 3.3. Sorption Selectivity and Thermodynamic Swelling Potential
during Ion Exchange

SBS sulfocationite					
Tetracycline-hydrogen			Tetraethylammonium-sodium		
K_{sw}	K	$\Delta\Phi_{sw}$ (kJ/mol)	K_{sw}	K	$\Delta\Phi_{sw}$ (kJ/mol)
1.5	518	52.3	1.7	1.8	19.3
2.4	195	77.4	2.6	1.8	15.6
3.9	138	79.6	5.5	2.1	15.0

The quantity $\Delta\Phi_{sw}$ is the part of the thermodynamic potential of ion
exchange that is due to the change in the hydrated state of the ionite.
Bearing in mind that any change in the hydrated parameter in the external
dilute solution for the same electrolyte is independent of what type of
ionite is being used, we may assume that $\Delta\Phi_{sw}$ reflects the changes in the
selectivity of the ion exchange due to using ionites with differing crosslink ratios. As we mentioned above, the sorption of organic ions is
usually accompanied by a contraction in the ionite. The $\Delta\Phi_{sw}$ for ion
exchange in these systems (the organic ion is the l-ion) is positive; i.e., the
dehydration usually reduces the selectivity of the sorption of the organic
ions. Swelling the ionite (or reducing the crosslink ratio in the ionite)
increases the thermodynamic swelling potential of the ion exchange, $\Delta\Phi_{sw}$.
This results in the low sorption selectivity of very swollen ionites that
is usually considered to be a chemical phenomenon when other methods of
thermodynamically analyzing ion exchange (e.g., swelling pressure) are
used[160,162]. However, in individual comparatively rare cases, $\Delta\Phi_{sw}$ for
organic ion-metal ion (or other small symmetric ion) systems falls as the
ionite's swelling factor is increased. This increases the sorption selectivity of the more selected ions. Both situations are illustrated in
Table 3.3, in which $\Delta\Phi_{sw}$ is given for an ionite state close to when the
ionite is completely filled by the organic counter-ion.

3.10. SELECTIVE SORPTION OF ORGANIC IONS

Organic ions are often sorbed by ion exchange very selectively. It is
necessary therefore to study the reasons for this high selectivity, to
model the sorption, and thus to predict the ionite structures that will
have high selectivities with respect to organic ions. This is because high
thermodynamic selectivity enables us to develop processes for separating
and purifying physiologically active substances.

The selectivity of sorption for organic ions has been observed to rise
as their structure becomes more complex, as can be seen from the sorption
of quaternary ammonium bases[204]. Going from a methyl to ethyl base
(or one with a more complex radical), given sorption on a sulfocationite in
competition with a univalent metal ion, leads to a significant growth in
both the ion-exchange constant and the selectivity. Sorption selectivity
is increased even more if the quaternary ammonium ion base contains a
benzyl or phenyl residue, and very high ion-exchange constants have been
observed for the selective sorption of antibiotic ions (Table 3.4).
However, these results were for specially selected or synthesized ionites.
In order to obtain a model that describes the selective sorption of organic
ions, the effect of the ionite structure on the selectivity must be
explained and the relationship between the thermodynamic potential and the
ion-exchange enthalpy and entropy established.

Table 3.4. Ion-Exchange Sorption Selectivity for Antibiotics

Antibiotic	Second counter-ion	Ionite	K	ΔH° (kJ/mol)	TΔS° (kJ/mol)
Erythromycin	Na	CPAC	44	18.4	27.6
Chlortetracycline	Na	SBS	124	7.1	18.4
Oxytetracycline	Na	SBS	110	-15.1	-3.2
Oxyglucocycline	H	KRS-4	40	-7.9	1.3
Morphocycline	H	KRS-4	39	0	8.8
Morphocycline	H	KRS-4T40	36	6.7	15.5
Penicillin	Cl	ASB-3	19	18.4	25.5
Novobiocin	Cl	FAF $K_{SW} = 2.2$	955	16.7	33.5
Novobiocin	Cl	AV-16 $K_{SW} = 4.5$	725	0	16.3
Novobiocin	Cl	AV-17 $K_{SW} = 3.3$	1820	0	18.4

The way the thermodynamic ion-exchange constant depends on the fraction of ionogenic monomer in the copolymer is shown in Figure 3.20 for the sorption of triethylbenzylammonium ions onto sulfocationites containing phenyl residues with no charged substituents. The ion-exchange constant falls as the of number ionogenic groups is reduced and then increases when additional aromatic groupings, able to interact with the organic ions, appear in the local environment of the ionogenic groups. Thus by comparing the role of the microenvironment of the charged groups of the counter-ion with that of the fixed ions it is possible to find an analogy between the influence of the non-ionogenic groupings around the counter-ions and those around the fixed ions on the sorption selectivity for organic ions. A systematic study of the sorption of the ions of quaternary ammonium ion bases onto sulfocationites confirmed the role of both the immediate environment of the charged counter-ion groupings and those of the charged fixed-ion groupings in the selectivity of the ion exchange (Table 3.5). The table also gives information about the relationship between the changes in energy and entropy during ion exchange. The thermodynamic functions were calculated using the well-known relations

$$\Delta\Phi° = -RT \ln K,$$
$$\Delta H° = RT^2 \frac{\partial \ln K}{\partial T},$$
$$\Delta S° = \frac{\Delta H° - \Delta\Phi°}{T}. \qquad (3.63)$$

The enthalpy changes were determined for the individual systems by direct calorimetry.

Consider first Table 3.5 without paying attention to the last two cationites, viz., the sulfonated Sephadexes. A comparison of the constants along a row shows that the ion-exchange constants for all three synthetic cationites grow with the structural complexity of the counterions. The increase in the constants when going from aliphatic residues in the quaternary ammonium bases to triethylbenzylammonium is most marked. A similar pattern can be seen in the columns of the table, and there is a growth in the constants from top to bottom for each of the organic ions. Moreover, the molecular sorbability of alcohols has been shown experi-

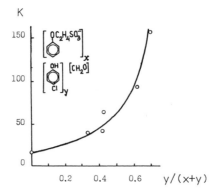

Fig. 3.20. Ion-exchange constants for triethylbenzylammonium sorption
            onto a sulfocationite versus fraction of ionogenic monomer
            in a copolymer.

mentally to grow in the same direction for the same ionites. Thus a
systematic investigation of the sorption of the ions of quaternary ammonium
bases indicates that an additional non-ionic interaction has a large effect
on the selectivity of ion exchange. The low ion-exchange constants for the
sorption of these bases on the sulfonated Sephadexes (they are close to
unity for Na-H exchange, too) become understandable from this viewpoint.
In fact, the Sephadexes are very good materials for gel chromatography pre-
cisely because they are hydrophilic and have weak interaction with organic
ions (particularly macromolecular proteins). Hence ion-exchange Sephadexes
have weak additional interactions with organic cations, and this results
in their demonstrated low sorption selectivities. Most other organic
ions also sorb with poor selectivities on ion-exchange Sephadexes
and biogels. An example is the poor sorption selectivity for oxytetracycline on sulfonated Sephadex in comparison to the very great
selectivity for the ions on the standard synthetic sulfocationites (Figure
3.21).

Table 3.5 shows that the high sorption selectivity for triethylbenzylammonium is accompanied by large entropy rises. As we shall see, all the
results concerning the thermodynamic potentials, enthalpies, and entropies
of ion exchange with organic ions require a series of models to explain
them, although they can all be covered by the overall ideas about the
coexistence of ionic and weak interactions between the organic counter-ions
and the ionite. For most of the antibiotics that can be sorbed onto
synthetic polymer ionites the selectivity is usually accompanied by positive entropy of ion exchange (Table 3.4).

The simplest interpretation of a positive entropy effect (a positive
change - a rise - in entropy) for selective ion-exchange seems to be that
it is due to an additional hydrophobic interaction between the ionite and
the counter-ion[205]. Indeed, it can be seen from Figure 3.22 that the
selective ion-exchange sorption of oxytetracycline onto Dowex-50 × 2
sulfocationite, which has positive entropy changes $\Delta S°$ for exchanges with
hydrogen in aqueous solutions, falls when methanol, ethanol, butanol and
urea, or sodium dodecyl sulfate are added to the solution. This corresponds to the model that describes counter-ion/ionite interactions as
being a set of ionic and additional hydrophobic ones. On the other hand,
the data in Table 3.6 demonstrate that changing an aqueous solution to a
water-methanol one reduces the sorption selectivity for Novocaine on
carboxylic cationites, reducing at the same time the positive entropy
effect of sorption. This contradicts the hypothesis that hydrophobic
interactions between the counter-ions and the ionite are responsible for
the sorption selectivity.

Table 3.5. Sorption Selectivity for Quaternary Ammonium Ion Bases on Network Sulfocationites

Cationite	Second counter-ion	$(CH_3)_4N^+$				$(CH_3)_3C_2H_4OHN^+$				$(C_2H_5)_3C_6H_5CH_2N^+$			
		K	$\Delta\Phi^\circ$	ΔH°	$T\Delta S^\circ$	K	$\Delta\Phi^\circ$	ΔH°	$T\Delta S^\circ$	K	$\Delta\Phi^\circ$	ΔH°	$T\Delta S^\circ$
Dowex 50 × 2	Na	1.26	-0.6	-1.7	-1.3	1.61	-1.3	-6.7	-5.4	4.8	-5.0	-5.0	0
KU-5A	Na	2.9	-2.5	-2.2	-2.9	3.64	-3.3	-8.4	-5.0	20.0	-8.8	-3.8	5.0
KU-1	Na^+	6.9	-4.6	-5.9	-1.3	9.1	-5.4	-11.3	-5.0	162.0	-13.4	10.9	24.3
Sulfonated Sephadex SE-25	H^+	4.2				4				4			
Sulfonated Sephadex SE-50	H^+	4				4				4			

Note. K is the ion-exchange constant; $\Delta\Phi$, ΔH, and $T\Delta S$ are expressed in kJ/mol.
Basic monomer units: Dowex 50 × 2, $-CH$ $-CH-$; KU-5A, $-CH-$

; sulfonated Sephadex SE-25, dextran; sulfonated Sephadex SE-50, dextran.

KU-1,

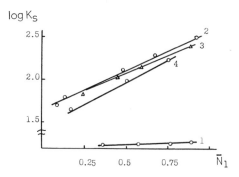

Fig. 3.21. Sorption selectivity for oxytetracycline on macronet
sulfocationites. Curves 1) sulfonated Sephadex SE-25;
and KMDM-6 with 2) 20%, 3) 30%, and 4) 10% HMDMA.

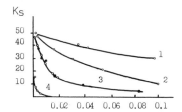

Fig. 3.22. Sorption selectivity for oxytetracycline on Dowex-50 × 2 in the
presence of competing ions. Curves 1) methanol; 2) ethanol;
3) butanol/urea; 4) sodium dodecyl sulfate.

To explain these and a number of similar results it is necessary to
introduce the idea that the ionite/organic counter-ion system has a set of
microstates while the ionite/mineral (small) ion system does not. One way
the microstates may arise is due to a hydrophobic effect, this being a
disruption in the quasi-crystalline aqueous structure by the hydrocarbon
residues and the acquisition by these water molecules of extra degrees of
freedom. The second alternative is the appearance of weak interactions
between individual groupings of the organic counter-ion and the structural
elements of the ionite. If the energy of the extra interaction is commensurate with the energy needed for thermal motion (kT), then each such
interaction adds another energy level in comparison with the interactions
between the ionite and small symmetric counter-ions. The greatest number
of microstates occurs when there is the greatest number of extra interaction points and energies close to kT.

The sorption selectivity for proteins on carboxylic cationites can be
seen by changing the number of ionogenic groups in the ionite, e.g., by
synthesizing the ionite with monomers that have no ionogenic groups, such
as hydroxypropylmethacrylamide, in addition to the methacrylic acid and
crosslinking agent (Figure 3.23). The selectivity is significantly raised
by increasing the volumetric concentration of carboxylic groups. It must
be assumed that the carboxylic groups are responsible for the additional
interactions. Probably what we have here are dipole-ion or dipole-dipole
interactions amongst a number of functional groups, thus creating a series
of new energy levels, i.e., a series of microstates.

Another explanation for sorption selectivity is illustrated in
Figure 3.24. Here an increase in the sorption selectivity for oxytetracycline can be seen for a series of sulfocationites that are sulfonated

Table 3.6. Ion-Exchange Constants for Antibiotics on Ionites Containing Divinylbenzene (DVB) or Divinyl (DV) as the Crosslinking Agent

Exchanging ions	Ionite	Crosslinking agent	K
	Anion exchange		
Penicillin-chloride	ASD-2	DVB	3.5
	ASB-2	DV	5.5
	ASD-3	DVB	11.2
	ASB-3	DV	22.0
	ASD-4	DVB	0.1
	ASB-4	DV	8.3
	Cation exchange		
Oxytetracycline-sodium	SDV-3	DVB	22.0
	SBS-3	DV	150.0

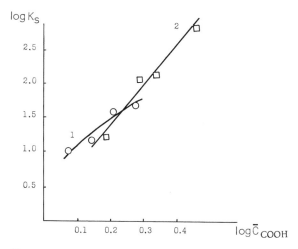

Fig. 3.23. Influence of concentration of carboxylic groups in Biocarb-T on the sorption selectivity for 1) hemoglobin and 2) insulin.

copolymers of styrene and divinylbenzene, telomerized in the presence of carbon tetrachloride. We know that telomerization is due to termination and transfer of growing chains. Thus a feature of a co-telomer is the large number of polymer chain ends; moreover, the number of these is raised by increasing the concentration of telogen (e.g., carbon tetrachloride). At the same time, the copolymer structure is altered. A network co-telomer has very high swelling factor, and its chains are very mobile. The growth in the organic-ion sorption selectivity as the ionite becomes more of a telomer must be due to the additional microstates that appear as a result of local weak interactions with either the polymer chain ends or the mobile sections of the chains.

In order to go from relatively unselective sorbents to very selective sorbents for organic ions, i.e., to obtain large changes in the thermodynamic standard potentials, there must be very significant restructuring of the network polyelectrolyte so as to enable the additional intermolecular interactions between the counter-ions and the ionite to form. Meanwhile it only needs small changes in ionite structure to cause very sharp changes, or even changes in sign, in the ion-exchange entropy or enthalpy. The $\Delta H°$ and $\Delta S°$ functions may also be changed from one counter-ion

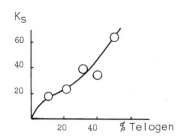

Fig. 3.24. Sorption selectivity for oxytetracycline on SDV-T sulfocationite versus telogen concentration (CCl_4) in copolymerization mixture.

to another. The compensation phenomenon for the ion exchange of hydrogen
and oxytetracycline ions on Dowex-50 sulfocationites with different crosslink ratios is given in Figure 3.25. All the systems are characterized by
high sorption selectivities for oxytetracycline. Moreover, the changes in
both the enthalpy and entropy go from negative values to positive ones.
Compensatory relations with transitions from negative to positive enthalpy
and entropy changes can also be traced for the exchange of quaternary
ammonium ion bases on the KU-1 sulfocationites (Figure 3.26). Finally, the
same pattern can seen for the sorption of the very complicated ions of
proteins onto carboxylic biosorbents (Figure 3.27).

The possibility of a compensation relation for a hydrophobic reaction
has already been appraised[205]. Here we shall dwell on the transition of
the $\Delta H°$ and $\Delta S°$ functions from negative to positive values for systems in
which a set of weak interactions at energies of kT leads to the appearance
of a series of microstates for the resinate and to an increased system
entropy. There is usually a steric hindrance when an organic ion approaches a fixed ion. Thus a positive change in entropy is naturally
accompanied by a positive change in enthalpy, due to which the energy of
the Coulomb interaction of the organic counter-ion with the ionite may be
less than the energy of the interaction between the small ion and the
ionite. However, given a limited change in the ionite structure, or for a
change to a counter-ion which interacts in the same way with the ionite, we
can expect the additional interaction for some or all of the groupings to
exceed kT significantly. The number of microstates for the resinate will
fall, reflecting the fall in the positive entropy change that occurs
because the counter-ions, which are bound to the ionite, have fewer degrees
of freedom. The Coulomb interactions between the organic ions and the
ionite and the additional interactions together have a larger energy than
the energy of the interaction between a small ion and the ionite. Thus a
change of sign for the enthalpy and entropy changes is not the exception
but the rule for ion exchange when organic ions are involved. Positive $\Delta S°$
are observed more often experimentally than the negative values. Since
ionic and additional (weak and hydrophobic) interactions coexist, we can
call them all polyfunctional interactions between organic counter-ions and
the ionite.

The number of microstates a resinate has may be calculated from
Boltzmann's equation, i.e.,

$$S = K \ln W = K \ln \frac{N!}{\pi N_i!} . \qquad (3.64)$$

We shall assume that a large organic ion can have n (e.g., ten) additional
bonds with the ionite. The number of energy levels, or microstates, will
thus, in the limit, be (n + 1) or significantly more for the independent

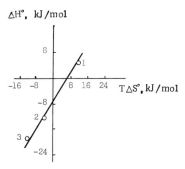

Fig. 3.25. Compensation effect for the exchange of hydrogen and oxytetracycline on Dowex-50 sulfocationites with differing crosslink
ratios. Points 1) 4.5%; 2) 1%; 3) 0.5% DVB.

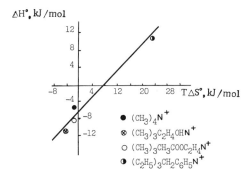

Fig. 3.26. Compensation effect for quaternary ammonium ion base sorption
on KU-1 sulfocationite with sodium as second counter-ion.

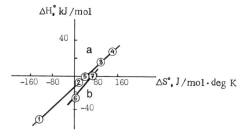

Fig. 3.27. Compensation effect for the sorption of proteins on Biocarb-T carboxylic cationite from a) water and b) methanol.
Points 1) thermolysin; 2) and 6) insulin; 3) ribonuclease;
4) blood albumin; 5) and 7) Novocaine.

microstates that might be formed as a result of any thermodynamic mobility
of the structural elements of the ionite. As a result, we can estimate that
the positive value of $T\Delta S°$ is 8-10 kJ/mol or more for the exchange of a
complicated ion for a small one.

There are grounds for suggesting that polyfunctional interactions
also occur between groups of organic counter-ions, giving rise to the cooperative effect that is seen in the growth of the sorption selectivity

as the mole fraction of the organic ion in the ionite is increased
(Figure 3.16). Another additional interaction that is a polyfunctional
interaction also contributes to the positive entropy change of the process.
Figure 3.28 shows how the sorption selectivity of a relatively small
organic ion grows with the increase in the mole fraction in the ionite,
and the growth of the positive enthalpy change, as determined by direct
calorimetry. The combination of a rise in enthalpy and a rise in selectivity means the entropy must rise too. Figure 3.29 shows the results of
a calorimetric investigation of the sorption of hemoglobin onto two types
of carboxylic cationite. Here, too, the enthalpy change is positive and
grows only when there is cooperative interaction, which in this case is
shown up by the growth in the first derivative of the sorption isotherm
with respect to concentration (selective sorption).

The results and models we have just covered are shown schematically in
Figure 3.30. The system of ideas we presented, including all the possible
polyfunctional variations, does not embrace all the microstates that can
appear in resinates. Ionite compaction during the sorption of organic ions
may be accompanied by the appearance of weak interactions between part of
the net structure, which leads to an increase in the system's entropy. A
significant increase in the strength of the interactions may have the
opposite effect of lowering the entropy because the mobility of the network's structure is impaired. The conformational components of the ion-

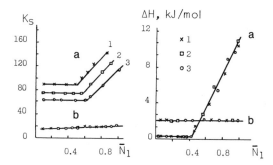

Fig. 3.28. Cooperative effect for the sorption of organic ions on
Biocarb-T with a) 0.1 N NaCl in water and b) 0.1 N NaI
in methanol. Points 1) Novocaine; 2) oleandomycin;
3) oxytetracycline.

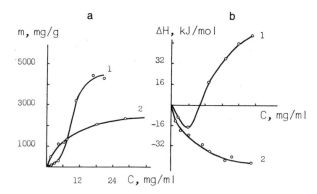

Fig. 3.29. Cooperative effect for sorption of hemoglobin by carboxylic
cationites. a) Equilibrium sorption isotherm; b) differential
heat of sorption. Curves 1) copolymer of methacrylic acid and
EDMA; 2) copolymer of acrylic acid and EDMA.

82

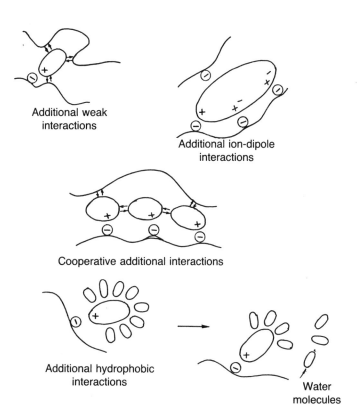

Additional weak
interactions

Additional ion-dipole
interactions

Cooperative additional interactions

Additional hydrophobic
interactions

Water
molecules

Fig. 3.30.   Schematic representation of polyfunctional interactions between
organic ions and ionites.

exchange entropy may have different signs and lead to either an increase or
a decrease in the sorption selectivity for organic ions.

An area that remains poorly studied quantitatively is the sorption of
polyvalent organic ions. It is known that the sorption of divalent metal
ions when exchanged for monovalent ions is accompanied by a rise in
entropy[206]. It has been proposed that two microstates arise for each
divalent ion compared with one microstate for a pair of the monovalent
ones. A number of stipulations have to be made to model this situation.
Firstly, it is necessary for the monovalent resinates to be identical, and
there must be equivalent ion-exchange. The divalent ion on the ionite must
have at least two energy microstates, for which a large energy of one bond
with a fixed ion and a low energy of the bond with another fixed ion, the
second being lower because the mobility of the ionite structure is limited,
are possible states. Even larger steric hindrances will occur between
polyvalent organic ions and the ionite. However, these systems have been
poorly studied, either calorimetrically or with respect to the ion-exchange
thermodynamic functions. Thus it is premature to suggest any model that
might describe them.

The polyfunctional model of a resinate with an organic counter-ion,
the schemes given in Figure 3.30, and the compensation relations demonstrate that resinates can form bonds with large energies with ionites. It
should be expected that the Coulomb interactions between organic counter-
ions and ionites are weaker, as a rule, than ionite interactions that
release a small ion during ion exchange. However, the set of additional
interactions may significantly exceed the ionite/small ion bond energy,

particularly if the energy of each local interaction is substantially greater than that needed for thermal motion and the number of bonds is large. Sorption such as this may lead to the organic ion becoming irreversibly bonded; i.e., desorption will be difficult even if many of the parameters of the external solution are changed. The reversibility of sorption is thus very complicated. Information about the kinetics of ion exchange is needed to estimate it, while to do this for column experiments dynamic and kinetic relationships and critical parameters that determine the degree of process completion are needed. These topics will be covered in Chapters 4 and 5.

3.11. SORPTION OF ORGANIC IONS AND IONITE PARTICLE SIZE

The permeability of networks was discussed in Chapter 2. It was suggested that the counter-ions keep to a limited layer close to the surface of the ionite particles because of the structural heterogeneity and the existence of canals and lightly crosslinked regions that only have limited dimensions. It has been suggested that for this reason and also because the network elements become more mobile when the ionite is fragmented the ionogenic groups will become more accessible for large ions if the particle dimensions are made smaller. This has been proved true in every case, and in particular the sorption capacity of a carboxylic cationite with respect to streptomycin is increased when smaller particles are used[199].

The kinetics of ion exchange is heavily dependent on ionite particle size. It should be noted that there are two independent factors that determine the rate at which equilibrium is established. On one hand, the radius of the particle determines the mass transfer rate due to a reduction in the diffusion path length. This is very important for gel diffusion, and mass transfer is accelerated by film diffusion too because of the increased contact surface area. In addition the increase in network mobility due to reduction in particle radius may also increase the gel diffusivity. The experimentally discovered possibility of going from gel to film (or mixed) diffusion for the sorption of the tetracycline group of antibiotics when the ionite particle diameter is reduced to 40-50 μm enables us to get very efficient ion exchange within a short period of time[207].

Naturally, we should expect particle dimension to have a large impact on the ability of the ionite to sorb large organic ions, particularly proteins[180,208]. Indeed, and as Figure 3.31 demonstrates, the sorption of blood albumin is increased by the grinding of both cationites and anionites. The limiting sorption capacity is about 6 g/g for particle radii of

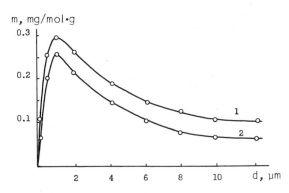

Fig. 3.31. Sorptive capacity of blood albumin versus ionite particle size on 1) AV-17 and 2) AV-16 anionites.

2-5 μm. Further grinding of the ionite leads to a certain fall in the
sorption of the protein. The curves show how both the accessibility of the
ionogenic groups and the sorption selectivity grow. The latter is especially clear in relation to the fall in the sorption capacity when the ionite
is intensively ground to particle dimensions of 1 μm or less. The thermodynamic result of the growth (and fall) in the sorption selectivity for
organic ions due to ionite grinding was ascertained experimentally by
studying the sorption of oxytetracycline onto the same sulfocationite
ground to different particle sizes. The differences in the particle
dimensions were obtained by mechanically and then chemically grinding the
ionite (Figure 3.32). As we go from the standard ionite to a ground one
with a particle diameter of 7 μm, there is an increase in the sorption
selectivity for the oxytetracycline, especially when the concentration of
the organic ion in the ionite is high. Further fragmenting of the ionite

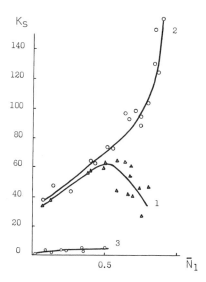

Fig. 3.32. Sorption selectivity for oxytetracycline on KRS-5P sulfocationite with different particle sizes. Average ionite
particle diameters: 1) 90 μm; 2) 7 μm; 3) 0.1 μm.

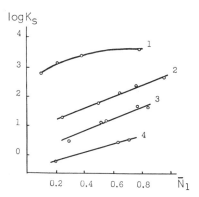

Fig. 3.33. Sorption selectivity for novobiocin on anionites with
differing particle sizes. Curves 1) and 4) ARA-3p-T40;
2) and 3) ARA-1p-T40. Average ionite particle diameters:
1) and 2) 10 μm; 3) and 4) 200 μm.

results in a fall in the oxytetracycline sorption capacity. Changes in the selectivity of the sorption of another antibiotic, novobiocin, confirmed the generality of the process (Figure 3.33).

The variability of sorption selectivity with respect to particle size must be interpreted from the general viewpoint of the selectivity for organic ions; i.e., polyfunctional interactions are the major factor in the high sorption selectivity for the organic ions. Dispersing an ionite or going to small radius beads, obtained from a direct synthesis of micro-globules, means that the ionite has new structural features. First of all, the network elements may become more mobile, which is seen both by the growth in the swelling factor, and from special investigations into the kinetic and thermodynamic mobilities of the polymer network, given that the crosslink ratio is maintained. Another factor is that the chemical and mechanical dispersion of the ionite creates more polymer chain-ends. It can be assumed that both factors increase the ability of the ionite to form additional bonds with the counter-ion. Very severe dispersion results in more sections being greatly swollen, and this must lower the number of sections of ionite that can form additional bonds. This argument follows in particular from a study of the way the sorption selectivity for oxy-

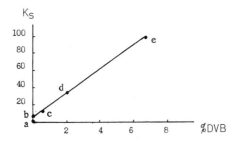

Fig. 3.34. Sorption selectivity for oxytetracycline versus crosslink ratio for Dowex sulfocationites (c, d, e) and a) polyvinyl-sulfonic acid and b) polysulfostyrene.

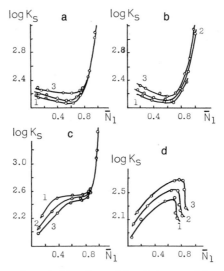

Fig. 3.35. Sorption selectivity for chlortetracycline by surface-layer cationites SK-6 (6% DVB). Thickness of sorbing layer: a) 3 μm, b) 6 μm, and c) 22 μm; d) is for Dowex 50 × 5. Temperatures: 1) 10°C; 2) 20°C; 3) 30°C.

86

tetracycline changes in sorption onto sulfocationites made with differing quantities of crosslinking agent (Figure 3.34). There is a sudden drop in selectivity when linear polyelectrolytes, a model of a network structure at the limit of dispersion, are used. The selectivity is altered by changes in the crosslink ratio and degree of swelling.

It is possible to compare ground ionites and surface-layer ionites. The latter are obtained by sulfonating a copolymer of styrene and divinyl-benzene that had been bound to impermeable beads (glass spheres). Surface-layer (Figure 3.35) and ground ionites have very high selectivities for the sorption of organic ions, especially when the mole fraction of the ions in the ionite is large.

3.12. ION EXCHANGE WITH PHYSIOLOGICALLY ACTIVE SUBSTANCES

The very wide use of ion-exchange sorption and chromatography for separating and analyzing physiologically active substances has generated a vast number of papers on the subject in recent times. Our presentation has meant we have not needed to mention work lacking quantitative aspects. In this section we shall dwell on experimental data that give the general features of ion-exchange equilibria involving organic counter-ions. Obviously, different classes of substances will have their own peculiar-ities, depending on the electrochemical properties of the sorbed ions, their molecular masses, and the relationships between the properties of the ionite and counter-ions and the properties of the solution.

3.12.1. Antibiotics

The most widely used medical antibiotics are organic substances with molecular masses in the hundreds of daltons. Consequently, most of them are electrolytes and can be considered as counter-ions. The way they interact with ionites shows that antibiotics can be compared both in terms of their ability to overcome the difficulties in moving through the ionite's network structure, and in terms of some of the general laws of ion exchange[8,11,15,209,210]. A very general picture of the behavior of organic counter-ions was obtained from a detailed analysis of thermodynamics of cation and anion exchange involving antibiotics on ionites with differing network densities. Network density, for gel ionites, is conveniently characterized by the swelling factor, and not by the crosslink ratio because both polymerized and polycondensed ionites have been studied (Table 3.7)[209]. The data in the table demonstrate that a rise or fall in the ion-exchange constants can be compared with the sign of the entropy or enthalpy component of the standard thermodynamic potential of ion exchange. A reduction in the ion-exchange constant is supposed to accompany a rise in the ionite's swelling factor, for example, when the crosslink ratio in the ionite is decreased[136]. This situation is modelled in a number of ways, some using the idea of a swelling pressure, others not. The important factor here is the ionite's compaction when an organic ion is sorbed. This corresponds to dehydration and a fall in the absolute value of the negative thermodynamic potential, which usually grows as the ionite's swelling factor increases. This is all characteristic of a system in which greater sorption selectivity is related to decreases in enthalpy. The inverse, i.e., growth in ion-exchange constant, is seen in systems for which greater selectivity depends on increases in entropy. One interpretation is that for processes with positive entropies for the sorption of an organic ion the counter-ion/ionite bond energy is small, and that the sorption itself does not lead to any significant dehydration of the ionite. The selectivity growth when the swelling factor is increased is thus a consequence of the greater mobility of the network's structural elements, which leads to more local additional bonding (an alignment of the functional groups in the ionite and counter-ion) with low bond energies.

Table 3.7. Ion-Exchange Constants for Antibiotics on Ionites with
           Differing Swelling Factors

Exchanging ions	Ionite	K_{sw}	K	$\Delta S°$	$\Delta H°$
Oxytetracycline-hydrogen	SBS	1.4	406	< 0	< 0
		2.4	344		
		5.1	188		
Oxytetracycline-hydrogen	Dowex 50	1.8	270	< 0	< 0
		2.8	90		
		3.7	58		
		9.3	28		
Erythromycin-sodium	CPAC	2.2	69	> 0	> 0
		3.2	83		
		4.6	95		
Penicillin-chloride	FAF	1.5	6.1	> 0	> 0
		3.4	7.8		
		3.4	8.6		
Chlortetracycline-sodium	SBS	3.0	100	> 0	> 0
		7.7	132		

This corresponds to an additional growth in the entropy of the system when
the dehydration factor is of limited importance. We shall use the idea
that there is a gap between a counter-ion and the fixed ion in the ionite
(for $\Delta S° > 0$) to explain the new effects that have been observed when
investigating the kinetic permeability of ionites to organic ions. As
regards the variability of solvation, a special study of a model in which
the ion-exchange constant rose for increases in the swelling factor (Table
3.8) demonstrated that an increase in the degree the ionite swelled did not
raise the positive thermodynamic potential of swelling; indeed, it rather
reduced it.

    3.12.1a. Antibiotics of the Tetracycline Group. Tetracycline, its
analogs, and chemical modifications are amphoteric compounds that can
participate in either cation or anion exchange. The best studied repre-
sentatives of this group, besides tetracycline (TC) itself, are oxytetracycline (OTC) and chlortetracycline (CTC) (Figure 3.36). Potentiometric
titration and spectral analysis studies of the ionization of individual
groups[211,212] demonstrate that in acid media they exist as cations, in
neutral solutions they are in the form of zwitterions, while in alkali
media they are anions. The ionization constants pK_{a1}, pK_{a2}, and pK_{a3} refer
to the enolic grouping (A), the phenolic diketone grouping (C), and the
protonated amino group (B), respectively. The dissociation of these compounds can be represented as

$$A\ B^+C \xrightleftharpoons{pK_{a1}} A^-B^+C \xrightleftharpoons{pK_{a2}} A^-B^+C^- \xrightleftharpoons{pK_{a3}} A^-B\ C^-$$

The sorption selectivity of these groups on sulfocationites is very great,
as we have mentioned before. This and their ability to change the number
and sign of their charges not only enables them to be separated and
purified by ion exchange, but also makes it possible to use them as a
unique model to establish the laws governing the sorption of organic ions.
There is a very large quantity of work on this topic[71,145,180,213-225].

    Most work has been done for pK_{a1}, i.e., for cation exchange in acid
media. Dynamic experiments for pH < pK_{a1} (pH 1.9) and for pH > pK_{a1} (pH 4)
in the original solutions showed that in the first case the ion exchange
was equivalent (Figure 3.37), calculated from the extra concentration of
released hydrogen ions. Meanwhile, at pH 4 the quantity of sorbed tetra-

Table 3.8. Sorption Selectivity and Thermodynamic Swelling Potential for Tetraethylammonium-Sodium Exchange on SBS

K_{sw}	K	$\Delta\Phi_{sw}$ (kJ/g-equ)
1.7	1.8	19.2
2.6	1.8	15.5
5.5	2.1	15.1

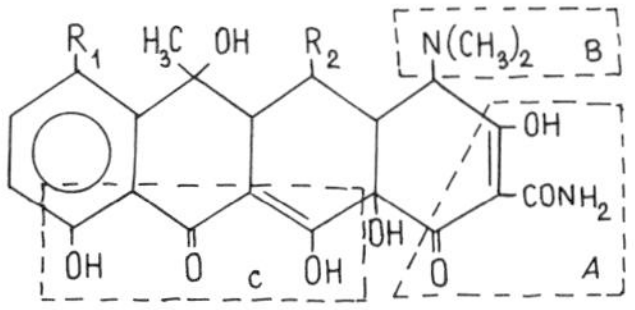

Fig. 3.36. Structure and properties of the tetracycline group.

TC: R_1 = H; R_2 = H; pK_{a1} = 3.30; pK_{a2} = 7.68; pK_{a3} = 9.69.

OTC: R_1 = H; R_2 = OH; pK_{a1} = 3.27; pK_{a2} = 7.32; pK_{a3} = 9.11.

CTC: R_1 = Cl; R_2 = H; pK_{a1} = 3.30; pK_{a2} = 7.44; pK_{a3} = 9.27.

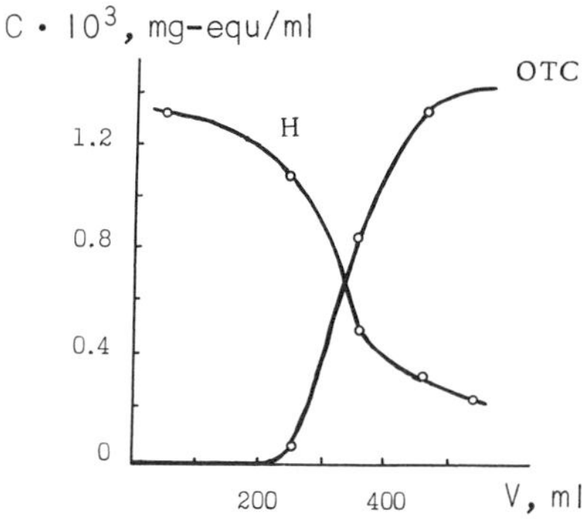

Fig. 3.37. Equivalence of ion exchange on SBS-3 sulfocationite involving oxytetracycline (OTC).

cycline considerably exceeded the quantity of desorbed hydrogen. It is possible therefore to assume that for pH > pK_{a1} the tetracycline is protonated during the exchange. This means its sorbability by cationites in static regimes is reduced by increases in the solution pH (Figure 3.38). The sorbability reduction in acid media corresponds to competition with sodium ions. However, sodium and tetracycline ions exchange equivalently in a wide pH interval (Figure 3.39), which indicates that they can form another sort of resinate including the tetracycline, probably in the form of a dipolar ion. Tetracycline group ion-exchange has been studied in this respect for pH $\ll$ pK_{a1}, for pH < 2 with hydrogen ions, and for pH 1.9 with sodium ions. In the latter case the concentration of sodium ions has to exceed the concentration of hydrogen ions significantly in order for the correction to the ternary system T^+ - Na^+ - H^+ to be insignificant.

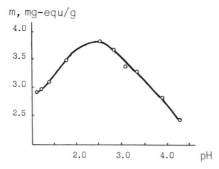

Fig. 3.38. Sorption capacity for tetracycline on SBS-3 sulfocationite versus solution pH.

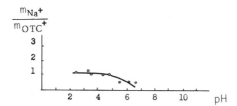

Fig. 3.39. Ion-exchange equivalence for oxytetracycline versus solution pH on SBS-3 sulfocationite. $\dfrac{m_{Na^+}}{m_{OTC^+}}$ is the ratio of the quantity of desorbed sodium to the quantity of antibiotic absorbed by the ionite (in mg-equ).

The thermodynamic functions for these (and most other) monovalent ion/monovalent ion exchange systems are determined from an analysis of the way the selectivity depends on the mole fraction of the organic ions over a wide range of loading factors of the ionite by the organic ions for several (three or four) temperatures. The ΔH° values obtained in this way have been compared with those calculated using calorimetry, and the two sets of data are satisfactorily close for most systems. The experimental data for the TC-H, OTC-H, and CTC-H systems are given in Table 3.9.

We considered the changes in sign in the ΔH° and ΔS° data earlier using the polyfunctional interaction model. Consider the negative values of ΔH° for the CTC-H system. This indicates a large interaction energy between the CTC and the ionites, and the ion-exchange constant of this system is higher than those of the TC-H and OTC-H systems. It should be noted that this sort of interaction is not desirable for practical purposes because it hinders the desorption of the ions even when high-pH solutions are used, which contrasts with the complete desorption of TC and OTC under the same conditions. KRS ionites, which are synthesized using para-divinylbenzene, have poorer selectivities for the whole tetracycline group than the analogous ionites synthesized using a mixture of divinylbenzene isomers. Apparently, the more regular structure of the KRS ionites reduces the number of polyfunctional interactions between the ionite and the organic ions (the tetracyclines in this case). Synthesis of the KRS ionites in the presence of telogen has no significant effect on the selectivity of the sorption of the tetracyclines, which is not the case when the DVB monomer mixture includes telogen. In the latter case ionites with much higher ion-exchange constants are obtained for the sorption of the

Table 3.9. Thermodynamic Functions for the Exchange of TC, OTC, and CTC with Hydrogen on KRS*

Cationite	K	$\Delta\Phi°$	$\Delta H°$	$T\Delta S°$
			kJ/g-equ	
		TC–H		
KRS-2	89	-10.9	-26.8	-15.9
KRS-4	200	-12.6	7.0	20.9
KRS-4T40	120	-4.7	0	11.7
		OTC–H		
KRS-2	53	-9.6	0	9.6
KRS-4	98	-11.3	-7.5	3.8
KRS-4T40	100	-11.3	4.2	18.0
		CTC–H		
KRS-2	150	-12.1	-7.5	4.6
KRS-4	270	-13.8	-12.6	1.3
KRS-4T40	230	-13.1	-19.2	-5.9

*KRS-2 and KRS-4 contain 2% and 4% DVB, and KRS-4T40 is obtained with 40% CCl_4 in the monomer mix.

tetracycline group (Figure 3.43). Matrix structure has an important role in the sorption selectivity for these antibiotics. Table 3.10 contains the sorption selectivities for OTC onto polycondensed and polymerized sulfo-cationites. Changes in the component ratios are given when the mole fraction of the OTC or CTC is less than 0.1. Increasing the crosslink ratio in Dowex-50 ionites leads to an increase in selectivity. The poly-condensed cationites KU-1, KU-5, and KU-6 yield particularly large selec-tivities. It is worth mentioning that polycondensed ionites often have high selectivities for organic ions[226,227]. This is apparently due to the appearance of polyfunctional interactions between the organic ions and aromatic residues in the neighborhood of the fixed ions, which appear due to the network heterogeneity arising when the polymer is condensed from oligomers.

The cooperative effect is very clearly expressed during the sorption of the tetracyclines by sulfocationites. The growth in K_s for the sulfo-styrene Dowex-50 occurs in almost every system of this type, starting once a quarter of the fixed groups has been filled by the ions (OTC in Figure 3.40). The fall in selectivity is due to structural heterogeneity, because all the active centers able to interact with large energies with the OTC counter-ions are exhausted and because the cooperative effect is absent for these centers. Sorption selectivity, over almost the whole range of counter-ion concentration in the ionite, rises as the crosslinking agent (divinylbenzene) concentration increases. When using the more regularly structured analogs of KRS, the cooperative effect is observed for the sorption of TC, OTC, and CTC (Figure 3.41). Moreover, energetic hetero-geneity of the ionite, seen as a fall in selectivity, appears only for CTC ions.

A study of selectivity for the tetracyclines enables us to compare monovalent-monovalent ion exchange with divalent-divalent exchange. In addition to anion exchange in alkali media[214], in which the antibiotics are very unstable and partially dissociated, cation exchange involving aminomethyl derivatives of tetracycline (morphocycline and oxyglucocycline) has been studied (Figure 3.42).

Table 3.10. Sorption Selectivity for Oxytetracycline on Sulfocationites with Differing Structures

Cationite	K_s (for $\bar{N}_1 < 0.1$)
Dowex 50 × 0.5	12
Dowex 50 × 1	15
Dowex 50 × 6	40
KU-6	250
KU-5	300
KU-1	1500

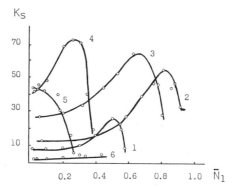

Fig. 3.40. Sorption selectivity for oxytetracycline-sodium versus mole fraction of organic counter-ion on Dowex-50. Curves 1) 0% DVB; 2) 0.5% DVB; 3) 2% DVB; 4) 6.6% DVB; 5) 8% DVB; while 6) is for polyvinylsulfonic acid.

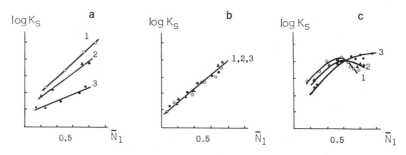

Fig. 3.41. Sorption selectivity for a) tetracycline, b) oxytetracycline, and c) chlortetracycline versus mole fraction of organic counter-ion on KRS-2 sulfocationite. Temperatures: 1) 10°C; 2) 20°C; 3) 30°C.

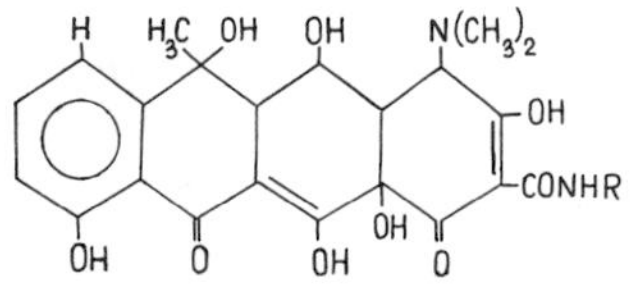

Fig. 3.42. Structures of morphocycline and oxyglucocycline.

Oxyglucocycline: R is $-CH_2-N(CH_3)-C_6H_{12}O_6$.
Morphocycline: R is $-CH_2-NC_4H_8O$.

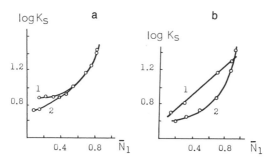

Fig. 3.43. Sorption selectivity for a) oxyglucocycline and b) morphocycline versus mole fraction of organic counter-ion in the
ionite. Curves 1) KRS-2 and 2) KRS-4T40; hydrogen was the
second counter-ion.

The patterns underlying the exchange of ions with different valences
are known to differ considerably for those underlying the exchange of
similar valence ions. Thus, diluting the solution - lowering the overall
concentration of electrolyte - leads to an increase in the relative absorption of monovalent ions. Hence comparisons of the ion-exchange constants
or selectivity coefficients for the exchange of ions of different valences
do not yield unambiguous information about the selectivity of an ionite.
Nonetheless, equivalent concentrations and the corresponding calculated
selectivities and ion-exchange constants can be used for relative information about an ionite's selectivity. The curves in Figure 3.43 show the
selectivities of the sorption of morphocycline and oxyglucocycline in
competition with hydrogen ions.

3.12.1b. Erythromycin and Oleandomycin. Studies of monovalent-
monovalent cation exchange involving these macrolide antibiotics can help
explain how resinates including these ions depend on the pH and ionic
strength of the external solution.

Erythromycin (EM) and oleandomycin (OM) are cyclic molecules with one
potential ionogenic group - dimethylamino (Figure 3.44). They have close
pK_a's, viz., 8.6 for EM and 8.5 for OM[228,229]. EM is particularly unstable in acid media and partially stable in alkalis. A convenient range
for studying ion exchange is pH 6-7. The antibiotics are almost completely
ionized in the solution in this range as singly charged cations. A feature
of the sorption of OM and EM ions onto cationites (especially carboxylic
ones) is the change in the ionization of the ionite when the ions are
introduced[230]. Thus during the sorption of OM and EM onto the sodium
forms of the cationites the real picture involves a three-component system
because the sodium and hydrogen ions must be considered together with the
organic ones. These systems are analyzed on the basis of the EM^+-Na^+ and
OM^+-Na^+ subsystems, allowing for the resinate of the third component; i.e.,
the overall number of resinates with organic counter-ions and sodium ions
was constant for many systems.

Sorption selectivity for OM and EM has been studied using both weak
and strong cationites[230,232-237], and on sulfocationites of differing
structures for initial solutions containing 1.5×10^{-3} mg-equ/ml of
antibiotic and 0.1 mg-equ/ml of sodium ions (Table 3.11).

EM is sorbed with more selectivity than OM. However, the standard
ionite, a sulfonated copolymer of styrene and divinylbenzene, does not sorb
these antibiotics very selectively. In order to sorb them there must be
neighborhoods close to each grouping that can have additional interactions

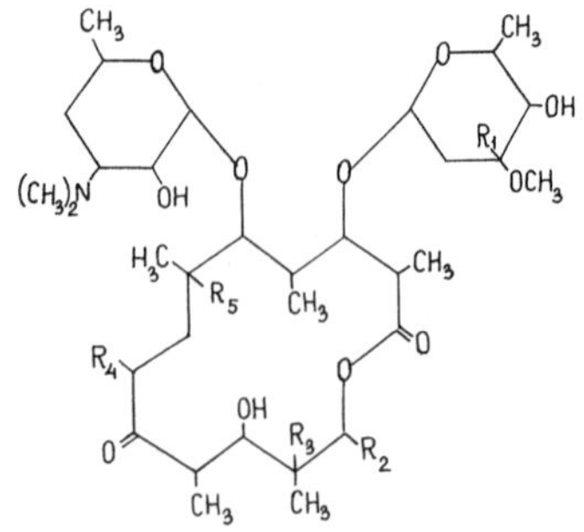

Fig. 3.44. Structures of erythromycin and oleandomycin.
Erythromycin: R_1 = CH_3; R_2 = CH_2-CH_3; R_3 = OH; R_4 = CH_3; R_5 = OH.
Oleandomycin: R_1 = H; R_2 = CH_3; R_3 = H; R_4 = ◁CH_2; R_5 = H.

Table 3.11. Sorption Selectivity for Erythromycin and Oleandomycin on Sulfocationites with pH 7. Sodium Was the Second Counter-ion

	Cationite		K_S	
Type	Ionogenic monomer	Crosslinking agent	EM	OM
SDV	Sulfostyrene	DVB	8	1
SBS	Sulfostyrene	DV	60	20
KU-5	Naphthalenesulfonic acid	Formaldehyde	80	30
SNK-D	MANS	DVB	120	60
SNK-E	MANS	EDMA	137	60
SNK-H	MANS	HMDMA	130	60

with the organic counter-ion. Naphthalene residues are the centers for the additional interactions in both polymerized and polycondensed sorbents, this being true incidentally for the sorption of many other organic ions.

However, in the case of the most specific SNK-type sorbents the selectivity of OM and EM sorption falls as the mole fraction of the counter-ion in the ionite is increased. This corresponds to a limitation in the way in which the organic ions can interact with each other (Figure 3.45) and is usually ascribed to the energetic heterogeneity of the ionite's functional groups. A thermodynamic analysis of these systems[237] demonstrates that both the EM and OM are sorbed at the expense of a rise in the system's entropy with a positive enthalpy term. Bearing in mind that for practical purposes antibiotics should be sorbed under conditions that ensure high selectivity and ensure that the ionite's functional groups are as heavily loaded as possible, it will be realized that OM and EM are very complicated materials for the development of preparative separation and purification.

Given that OM and EM are relatively strong bases, they may be sorbed onto carboxylic ionites, as well as onto sulfocationites. As was the case with the latter, the standard carboxylic ionites sorb the antibiotics with poor selectivity. Thus the methacrylic and acrylic acid copolymers with divinylbenzene have selectivities less than 10, and even the polycondensed CPA cationite, which contains phenoxyacetic acid (Figure 3.46), sorbs them with low selectivity. However, the addition to the copolymer of a hydrophilic monomer that does not contain an ionizable grouping (e.g., para-

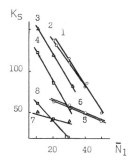

Fig. 3.45. Influence of mole fraction of organic counter-ion on
sorption selectivity for erythromycin (1, 2, 3, 4) and for
oleandomycin (5, 6, 7, 8). Lines 1) and 5) SNK-E; 2) and
6) SNK-H; 3) and 7) SNK-D; 4) and 8) SNK-M.

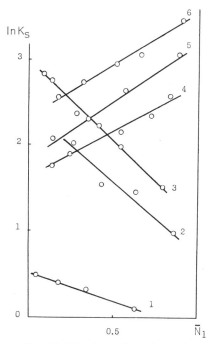

Fig. 3.46. Sorption selectivity for oleandomycin versus mole fraction
of organic counter-ion in carboxylic cationites with
different structures. Lines 1) CPA; 2) KFFU; 3) CPAC;
4) KRFFU; 5) KRFU-4; 6) KRFU-8.

chlorophenol or resorcinol) increases the selectivity dramatically (Figure
3.46). Ionites made by condensing phenoxyacetic acid with resorcinol and
formaldehyde have some important properties. It is difficult to ascribe
the cooperative effect that appears in the case of OM to interactions
between the ions because the exchange capacity of the ionite, the ability
of the molecules to associate, and other types of weak interaction are all
small. Apparently the organic ion causes conformational changes in the
ionite's network when it is sorbed, resulting in stronger interactions
between the sorbed ion and neighboring active centers (Table 3.12).

Table 3.12. Thermodynamic Functions for the Exchange of Oleandomycin for Sodium Ions on Carboxylic Ionites at pH 7

Ionite	ln K	$\Delta\Phi°$	$\Delta H°$	$T\Delta S°$
			kJ/g-equ	
CPAC	2.08	−5.0	35.1	40.2
KRFU	2.52	−6.3	18.8	25.1
KRFFU	2.08	−5.0	41.8	46.9

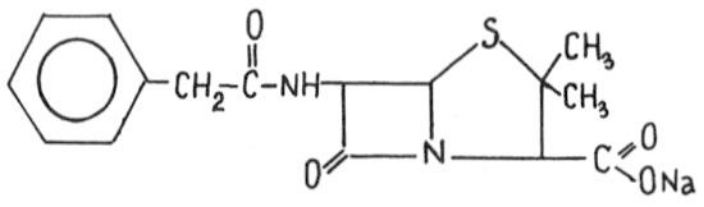

Fig. 3.47. Structures of benzylpenicillin and novobiocin.

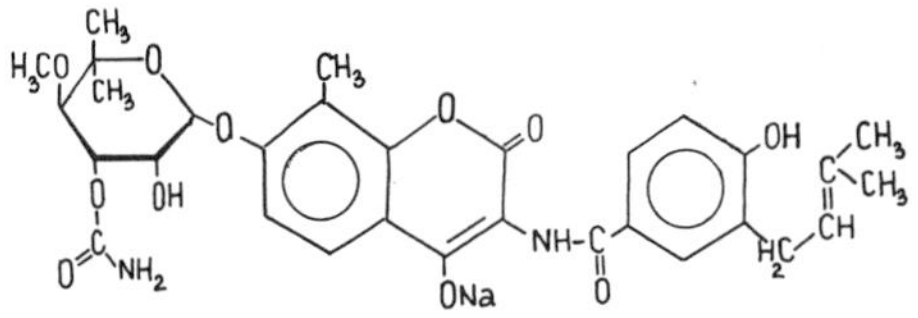

3.12.1c. Penicillin and Novobiocin. The anion exchange of organic ions has not been studied as much as cation exchange. Of the antibiotics with acidic properties, penicillin[239-245] and novobiocin[246-251] have been used the most for quantitative studies of their sorption onto anionites. The structures of these antibiotics are shown in Figure 3.47; benzylpenicillin has one acidic grouping with a pK_a of 2.7 (phenoxypenicillin has similar properties), while novobiocin contains two (pK_{a1} = 4.3 and pK_{a2} = 9.1). Both antibiotics exist as singly charged anions in neutral solutions, and in such solutions it has been found that they are involved in monovalent/monovalent anion exchange on a variety of ionites. However, the anion exchange of penicillin is significantly different from that of novobiocin.

Benzylpenicillin sorption is different firstly in that the specificity of bonding to the ionite is higher in the sulfate or phosphate form of the ionite than in the chloride form, and is greater when either sulfate or phosphate salts are added to the solution. Moreover, it is extremely difficult to desorb penicillin (particularly benzylpenicillin) from an ionite in the presence of phosphates or sulfates. Thus attempts to use many anionites to separate benzylpenicillin from native solutions have not obtained good yields, though initially precipitating polybasic acids using barium salts during sorption experiments enables all the benzylpenicillin to be extracted. Note that the anions of sulfuric and phosphoric acids react with benzylpenicillin on ionites to form stable complexes which cannot be disrupted by the limited changes in pH that are permissible when working with penicillin or when using aqueous organic electrolyte solutions. Anion exchange involving benzylpenicillin has been studied for exchange with chloride ions. The exchange constants for these systems are

Table 3.13. Thermodynamic Functions for Exchange of Novobiocin and
Chloride on Anionites

Anionite	K	$\Delta\Phi^\circ$	ΔH°	$T\Delta S^\circ$
			kJ/g-equ	
FAF	955	−16.7	18.7	35.1
FAKh	630	−15.5	20.9	36.4
AV−16	725	−16.3	0	16.3
AV−17	1820	−18.4	0	18.4

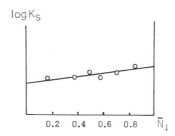

Fig. 3.48. Sorption selectivity for novobiocin on FAF anionite versus
mole fraction of organic counter-ion on ionite.

not large for either polymerized or polycondensed ionites[240,243,244].
The process takes place with an increase in entropy. Spectroscopic investigations have shown that the additional interactions that can occur in
these systems involve hydrogen bonds[122].

The interactions between novobiocin and ionites are very selective.
Moreover, while the tetracyclines are only sorbed very selectively by
ionites with a certain structure, novobiocin is very selectively sorbed by
a wide variety of densely crosslinked ionites (Tables 3.13). Novobiocin's
selectivity for all the systems studied is due to the growth in entropy.

The specific interactions between the novobiocin and the ionites,
which are characterized by a rise in the system entropy, are different from
the cooperative effect, which is the growth in selectivity accompanying a
rise in mole fraction of the counter-ions in the ionite (Figure 3.48).

3.12.1d. Aminocyclitols (Streptomycin, Kanamycin, and Neomycin).
These antibiotics are aminoglucosides and polyvalent bases. They are
usually found in the form of a number of isomers and stereoisomers after
a biosynthesis, and there are a large number of chemical modifications of
them. Their ion exchange has mainly been studied for streptomycin A (which
we shall simply call streptomycin), the two unseparated neomycin stereoisomers B and C (to which all the "neomycin" data refer), and kanamycin.
Streptomycin, which has a molecular formula of $C_{21}H_{39}N_7O_{12}$, is a trivalent
base with pK_a's of 8.5, 11.5, and 12.1. Kanamycin ($C_{18}H_{36}N_4O_{11}$) is a
tetravalent base all of whose ionogenic groups are ionized within a narrow
range of pH having pK_a's of 6.3, 7.3, 7.8, and 8.2. All six ionization
constants of neomycin are close, being between 6 and 9.

The sorption of this group has mainly been studied for carboxylic
cationites, but aluminosilicate sorbents have been looked at too[15,57,
252-267]. One reason for this choice of ionites and the limited use of the

sulfocationites is the very high selectivity of the carboxylic cationites, their poor sorption capacities for weak bases, but their large absorption capacities for relatively strong bases. Another reason for preferring carboxylic cationites to sulfocationites is that it is difficult to desorb the antibiotics (particularly streptomycin) from the latter. This is because alkaline solutions deactivate the antibiotics, while acid solutions only work with the carboxylic cationites, because they reduce their ionization levels.

As a rule, all or most of the functional groups interact with the organic ions. Thus the streptomycin releases three sodium ions from the carboxylic or aluminosilicate (Permutit) ionites. The sorption selectivity for streptomycin by carboxylic cationites falls as the ionite is swollen. Neomycin sorption is characterized by the cooperative effect (Figure 3.49), while its selectivity coefficient K_S poorly reflects how selectively it is sorbed. In order to obtain low ion fractions of neomycin in the KMDM-6 carboxylic cationite, i.e., around 0.1-0.2, it is necessary to use solutions with neomycin concentrations of about 5×10^{-4} N and sodium concentrations of 1 N.

Polyvalent interactions appear even when the sorption capacity for polyvalent ions in competition with lower valence ions is increased when the solution is diluted (Table 3.2). The inaccessibility of some of the functional groups of an ionite for organic ions is not just associated with the presence of polymer "wall" sections. During the sorption of some complex ions, particularly multifunctional ones like streptomycin, an ion can cover unused fixed groups. This hinders other large ions approaching the unused groups. Adding hydrophilic monomers that have no ion-exchange groups to the synthesis mixture of the ionite, and thus reducing the ionite's exchange capacity, increases the ability of a carboxylic cationite to exchange all its sodium ions for the streptomycin[57]. The permeability of carboxylic cationites with respect to these antibiotics can also be increased by using acrylic acid copolymers instead of methacrylic ones[267].

Pseudo-equilibria can appear during the sorption of streptomycin and some other organic ions. These occur because the streptomycin sorbs rapidly onto only some of the active centers, after which the sorption rate decreases. The ionite can be made to accept more ions by grinding the ionite, raising the temperature, or altering the ionic strength of the solution. This phenomenon has received much attention and has been carefully studied[263,264,266]. These results should not be used to criticize

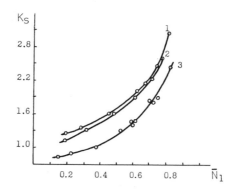

Fig. 3.49. Sorption selectivity for neomycin on KMDM-6 carboxylic
cationites versus mole fraction of organic counter-ion on
ionite. Curves 1) 30%; 2) 21%; 3) 12.5% HMDMA.

thermodynamic analyses of ion-exchange systems with limited accessibilities
for the organic ions because, firstly, these systems are the most important
practically and, secondly, the analysis can be done even when there are
small changes in the inaccessibility for large ions. It should be remembered that very many systems involving organic substances, as well as those
in the solid state, can be the subjects of reliable thermodynamic analyses,
even without the possibility of removing all their limitations and considering them as systems in metastable states.

3.12.2. Amino Acids and Proteins

Amino acids, polypeptides, and proteins are the best known electrolytes capable of existing in solution as both cations (in acidic media) and
anions (in alkaline media) and as zwitterions (these having both positive
and negative charges) in intermediate pH's. A fact of great importance for
the sorption of amino acids onto ionites is that monoaminomonocarboxylic
acids exist as cations and not zwitterions when sorbed onto sulfocationites
from neutral solutions[183]. Later spectral analyses confirmed that the
resinates of amino acids were similar[268]. It should be noted that the
cation form of monoaminomonocarboxylic acids bonded to a sulfocationite is
extended until all the ionite's ionogenic groups have been loaded with
organic ions. Thus it is not possible to consider the resinate states
solely from the point of view of local concentrations of hydrogen ions in
the ionite, even though the external solution's pH and the counter-ion
ratios in the ionite influence the ion-exchange constants. Investigations
of the equilibrium sorption of amino acids by ionites have been carried out
in many scientific centers[9,189,269-285].

The existence of amino-acid resinates in the form of sorbed cations on
cationites is a precondition of using the equations of equivalent ion-
exchange to analyze equilibria over a wide range of pH's and taking into
account the ionization. It is possible to consider the activity coefficients of the resinates of a number of very simple amino acids to be
relatively constant because the selectivity coefficients change slowly as
the concentration of sorbed amino acid builds up. However, the selectivity
coefficients of more complex dipole ions, used in the range of valid
equivalent ion-exchange isotherms, turn out to depend on the pH of the
external solution, as well as on the composition of counter-ions in the
ionite. One way of overcoming this difficulty in quantitative studies of
amino acid ion-exchange is to introduce dissociation concepts. For
example, it can be hypothesized that the dipole ions of amino acids are
sorbed together with cations, which would mean the sorption would depend on
the pH of the gel phase, and not just on that of the external solution. A
second approach is to assume that the activity coefficients of the resinates are varied by alterations in the pH's of the external and internal
solutions, the variations perhaps being significant. The most justifiable
quantitative study was done for two pH ranges, viz., for $pH < pK_{a1}$, when
the monoaminomonocarboxylic acids exist as cations in solution, and for
$pH \sim pI$, i.e., for the sorption of dipoles carrying equal numbers of
positive and negative charges.

The sorption of monoaminomonocarboxylic acids (and their electrochemically analogous polypeptides) in neutral pH's for hydrogen form
sulfocationites may be analyzed thermodynamically using different molecular
forms in neighboring phases, i.e., the dipole ion in the solution, a resinate including the amino acid's cation in the ionite phase, and the resinate
in the hydrogen form. The equilibrium is thus

$$\bar{\mu}_{R\pm} + \bar{\mu}_{H^+} - \bar{\mu}_{R^+} = 0, \qquad (3.65)$$

where $\bar{\mu}_{R\pm}$ is the chemical potential of the dipole ion in the solution,

$\bar{\mu}_{H^+}$ is the chemical potential of the resinate in the hydrogen form, and

$\bar{\mu}_{R^+}$ is the chemical potential of the resinate with the acid's cation.

By calling m_{R^+} and m_{H^+} the resinate concentrations, and $C_{R\pm}$ the zwitterion concentration, and using the corresponding activity coefficients $\bar{\gamma}_{R^+}$, $\bar{\gamma}_{H^+}$, and $\bar{\gamma}_{R\pm}$ we get

$$\frac{m_{H^+}C_{R\pm}}{m_{R^+}} = \frac{\bar{\gamma}_{R^+}}{\bar{\gamma}_{H^+} + \bar{\gamma}_{R\pm}} \, e^{-\Delta\Phi^\circ/RT} = Kf(\gamma). \qquad (3.66)$$

Then by assuming that $m_{R^+} + m_{H^+} = m$ is the overall exchange capacity of the ionite, we can transform Equation (3.66) into

$$m_{R^+} = \frac{mC_{R\pm}}{Kf(\gamma) + C_{R\pm}}, \qquad (3.67)$$

which is a Langmuir isotherm with saturation and corresponds to the experimental isotherm of the sorption of a monoaminomonocarboxylic acid (Figure 3.50).

The enthalpy and entropy components of the sorption of the amino acids can be found from the way the isotherms depend on temperature (Table 3.14). As was the case for equivalent ion-exchange involving organic ions, the direct sorption of dipole ions, which requires the transformation of the resinate from the hydrogen form to the amino acid form, is accompanied by a rise in the system's entropy.

Ion exchange in acid media (for pH < pK_{al}) has been studied for several amino acids and polypeptides using the isotherms of equivalent

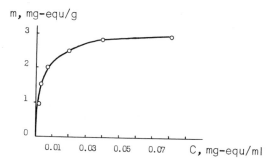

Fig. 3.50. Sorption isotherm for alanine on SBS-2 sulfocationite. K_{sw} = 2, pH = 7.

Table 3.14. Thermodynamic Functions of Sorption of Dipolar Ions onto SDV-3 (pH = 7)

Dipole ion	$\Delta\Phi^\circ$	ΔH°	$T\Delta S^\circ$
	kJ/g-equ		
Alanine	−13.4	0	+13.4
Glycylleucine	−13.8	0	+13.8
Glycylleucylvaline	−12.6	0	+12.6

ion-exchange. Figure 3.51 shows how the selectivity constants depend on the mole fractions of glycine and the dipeptide glycylphenylalanine in the ionite. The selectivity for the amino acid is small and is insignificantly changed by rises in the concentration of the counter-ion in the ionite, which means the system can be considered quasi-ideal, a conclusion we had reached before because the counter-ion is small and can be oriented around a fixed ion. The situation is completely the reverse for the dipeptide. Additional (polyfunctional) interactions show up here both in the rise in the selectivity and in the cooperative effect. In contrast, during the exchange of phenylalanine onto polycondensed hydrogen form sulfocationites the selectivity falls as the ionite gets filled up by the counter-ion. This indicates that the energetic heterogeneity of the ionite predominates over the possible cooperative effect (Figure 3.52).

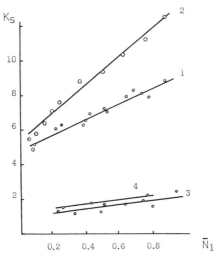

Fig. 3.51. Sorption selectivity for glycylphenylalanine (1, 2) and glycine (3, 4) on Dowex-50 sulfocationite versus mole fraction of organic counter-ion on ionite. Lines 1) and 3) 1% DVB; 2) and 4) 5% DVB.

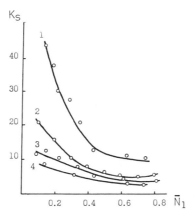

Fig. 3.52. Sorption selectivity for phenylalanine versus mole fraction of organic counter-ion in ionite on sulfocationites of different structures. Curves 1) KU-5; 2) KU-6; 3) KKB; 4) KFS.

A growth in the selectivity as more-complex amino acids, particularly aromatic ones, and peptides are sorbed is a tendency that is seen for many organic electrolyte systems. We would expect that during the sorption of dipolar ions polyfunctionality will appear with ion-ion and ion-dipole interactions[277,278]. The distinctive aspect of amino acid/hydrogen ion exchange for pH < pK_{a1} is the drop in entropy (Table 3.15), which can be seen as yet another aspect of the limited ability of the counter-ion to have additional interactions with ionites. A small growth in entropy is observed for phenylalanine on the KU-6 polycondensed sulfocationite. Comparing the sorption of amino acids for pH < pK_{a1} and for pH $\cong$ pI shows that the amino-acid resinates are energetically different for the different states.

The ion-exchange sorption of proteins is carried out on highly permeable ionites, the structures of which we considered in Chapter 2. The ionites with the highest protein selectivities incorporate a variety of polar and hydrophilic groups in their matrices and include copolymers of methacrylic (sometimes acrylic) acid with the hexahydro-2,3,5-triacryloyl-triazine, polymethylenebismethacrylamide, or ethylene glycol dimethacrylate as the crosslinking agent. In contrast, the ion-exchange Sephadexes cannot form additional interactions with organic ions, and sorb proteins with low selectivity. This is seen as a drop in the sorption capacity of the Sephadexes for proteins when considerable quantities of mineral salts are added. An important factor in the reversibility of protein sorption is the structural stability of the ionite's network, which is due to the insensitivity of the swelling factor to changes in the pH or ionic strength of the solution.

Proteins are amphoteric and can be modelled electrochemically by amino acids. However, their isoelectric points can be changed over a wide range, from pI $\sim$ 1 to pI $\sim$ 11 for some well-known proteins. Considering the sorption mechanism we described for dipolar ions and the fact that a monoaminomonocarboxylic acid resinate contains a cation, it is natural to assume that the effect depends on the distance between the amine and carboxy groups and that it may not be seen for proteins. In fact, the experimental evidence for proteins interacting with hydrogen ions and the sodium forms of the cationites can be interpreted in terms of the distance between the positively and negatively charged groups on the protein[9]. However, the state of protein resinates has not been looked at in detail; hence the models being offered for the sorption of proteins by ionites should be regarded with a certain degree of care.

Most of the work done on protein sorption has been for carboxylic cationites[115,116,286-292]. As opposed to sulfocationites, the carboxylic ionites have a low catalytic effect, on one hand (particularly when hydrogen ions are used as counter-ions), and, on the other, their ionizations can be regulated, thus giving control over sorption and desorption selectivity for proteins.

Table 3.15. Thermodynamic Functions (kJ/g-equ) for the Exchange of α-Amino Acids with Hydrogen for pH < pK_{a1}

Ionite	Glycine			Phenylalanine		
	$\Delta\Phi°$	$\Delta H°$	$T\Delta S°$	$\Delta\Phi°$	$\Delta H°$	$T\Delta S°$
Dowex 50 × 1	-1.2	-5.4	-4.2	-4.2	-9.2	-5.0
Dowex 50 × 5	-1.3	-6.7	-5.4	-5.4	-10.0	-4.6
KU-6		-		-5.0	-4.2	+0.8

In order to analyze how the sorption capacity of the ionites for proteins depends on the solution pH, we can use the relationships we derived in this chapter for weak electrolytes, assuming that only the cation form of the protein takes part in the establishment of the interphase equilibrium. We have already shown that the dependences of the sorption capacity on the solution pH are bell-shaped curves with maxima for this model if we can assume that the hydrogen un-ionized resinates of the carboxylic cationites are not involved in the sorption. The same sort of curves also come from more-detailed models constructed assuming that the interactions between the proteins and the ionites are combinations of ionic and dipole-ion ones[16]. The experimental data on the sorption of a variety of proteins onto highly permeable heteronet carboxylic biosorbents are given in Figure 3.53. The position of the maxima of these curves correlates with the isoelectric points of the proteins.

The complexity of the systems hampers any quantitative analyses of the sorption of proteins by carboxylic cationites. A consideration of the process for pH's corresponding to the sorption maximum of a protein is more reliable; i.e., it seems more rational to use Equation (3.67) in this case because the equation does not take into account the release of small counter-ions. The thermodynamic functions for the sorption of proteins by highly permeable carboxylic biosorbents can be found from the sorption isotherms and their temperature dependences (Table 3.16).

In most cases protein sorption is accompanied by a rise in entropy, which indicates that there are polyfunctional interactions with the ionite. Proteins should have clear hydrophobic interactions, and this is probably seen in a number of systems. However, the growth in the selectivity for proteins that is observed as the volumetric concentration of carboxylic groupings in the ionite is increased indicates that a blend of ionic and ion-dipole interactions is the most probable.

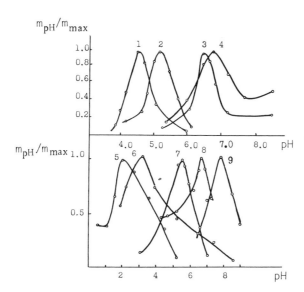

Fig. 3.53. Relative exchange capacities for the sorption of proteins on the carboxylic cationite Biocarb-T versus pH. Curves 1) terrilytin; 2) insulin; 3) chymotripsinogen; 4) RNase (pancreatic); 5) pepsin; 6) thymarin; 7) thermolysin; 8) hemoglobin; 9) lysozyme.

As we noted before, protein sorption by a carboxylic ionite is also dependent on the ionite's structure and ability to have weak interactions with the proteins. Figure 3.54 demonstrates the large, comparable absorption capacities of the sorbents Biocarb-T and Spheron (Separon) with respect to insulin, and the high selectivity of the sorption of insulin onto Biocarb-T.

Sulfocationites, both in the hydrogen form and in a composite form that contains some hydrogen ions, catalyze several reactions. Their hydrolytic effect on peptide bonds limits their use for sorbing and separating proteins. However, this effect is either absent or very weak for many sulfocationites with respect to insulin, and the sorption of insulin by sulfocationites has been studied in great detail[293-296]. In spite of insulin's relatively low isoelectric point, gel (SDV-T) and macroporous (KU-23) sulfocationites have a large sorption capacity for

Table 3.16. Thermodynamic Functions for the Sorption of Proteins onto the Heteronet Biosorbent Biocarb-T

Protein	$\Delta\Phi°$	$\Delta H°$	$T\Delta S°$
		kJ/mol	
Ribonuclease	-3.3	-2.0	1.3
Blood albumin	-4.0	5.4	9.4
Terrilytin	-5.5	0	5.5
Thermolysin	-54.0	-12.9	41.1
Insulin	-5.3	-4.4	0.9
Pepsin	-4.5	4.9	9.4

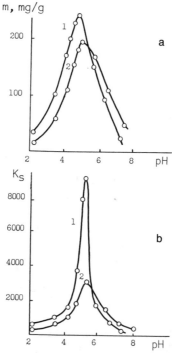

Fig. 3.54. Sorption of insulin by 1) Biocarb-T and 2) Spheron S-1000 versus pH. a) Sorption capacity; b) selectivity.

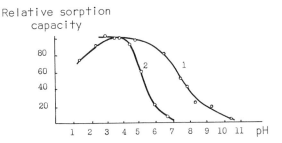

Fig. 3.55. Influence of pH on insulin sorption by 1) SDV-T gel and 2) KU-23 macroporous sulfocationites.

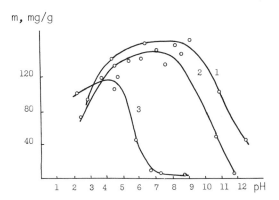

Fig. 3.56. Influence of pH on lysozyme sorption by KU-23 macroporous sulfocationite. Curves 1) 0.01 N; 2) 0.1 N; 3) 1 N NaCl.

insulin, especially at pH $\sim$ 4 (Figure 3.55). The right-hand end of insulin's sorption capacity versus pH curve falls more rapidly in a neutral pH region for the macroporous ionites than it does for the gel ionites, for which the sorption region extends to pH 10–11. For this reason, and due to better ion-exchange kinetics, the desorption of insulin from macroporous ionites is much simpler. The rigid structure of the macroporous sulfocationites means that the probabilities of both irreversible protein sorption and hydrolytic sorbent activity are less than those for gel ionites. Thus proteins other than insulin can be reversibly sorbed. Figure 3.56 gives the sorption capacity versus pH curve for lysozyme on the macroporous sulfocationite KU-23. Increasing the solution's ionic strength moves the sorption maximum and the right-hand section of the sorption capacity versus pH curve into an area that is more convenient for experimentation.

3.12.3. Nucleosides, Nucleotides, Alkaloids, Sulfonamides, and Miscellaneous Physiologically Active Substances

Although a vast number of publications about the use of ionites for sorbing, separating, and purifying a wide variety of physiologically active substances have appeared, very few have dealt with the patterns underlying their interactions with network polyelectrolytes. Thus in addition to the previous sections we shall present some of the quantitative patterns that can be seen in the interactions between ionites (both cationites and anionites) and organic ions (cations, anions, and amphoteric compounds).

Nucleosides are sorbed with low selectivity by the standard sulfonated copolymer of styrene and divinylbenzene (Figure 3.57). In contrast, SNK sulfocationites, which include naphtholsulfonic acid residues, sorb them with high selectivity[297,298]. A similar role can be seen to be played by naphthalene derivatives in ionite structures when the selectivities for adenosine (pK_{a1} = 3.5) sorption by polycondensed sulfocationites are compared (Table 3.17).

A thermodynamic analysis of the ion exchange of adenosine on very specific ionites showed that the adenosine was sorbed with a drop in entropy (Table 3.18). This should be interpreted as a consequence of strong additional interactions between the counter-ions and the ionites, with energies exceeding kT. The selectivity here falls somewhat if the swelling factor is reduced, i.e., if the crosslink ratio in the ionite is increased. It should be assumed that increases in network mobility due to the increase in the swelling factor are important if strong additional interactions between the ionites and the counter-ions are to be attained.

Nucleotides are amphoteric with $pK_{a1} \sim 1$, $pK_{a2} \sim 4$, and $pK_{a3} \sim 6$, the first and third constants being for the ionization of the phosphate groups. Thus nucleotides exist as singly charged cations for pH < 1, as dipolar ions (zwitterions) for pH 1-4, as singly charged anions for pH 4-6, and as

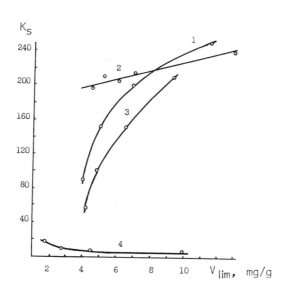

Fig. 3.57. Sorption selectivity for adenosine versus swelling factor on sulfocationites. Curves 1) SNK-E; 2) SNK-D; 3) SNK-H; 4) KRS.

Table 3.17. Sorption Selectivity for Adenosine on Polycondensed Sulfocationites

Ionite	Monomer	K_{sw}	$K_s(\bar{N}_1 < 0.1)$
KFS	β-Phenoxyethanesulfonic acid	3.5	15
KFSS	β-Phenoxyethanesulfosalicylic acid	3.3	17
KFSN	β-Naphthol and β-phenoxyethanesulfonic acid	3.1	75
KNS	β-(2-Naphthoxy)ethanesulfonic acid	3.5	80
KU-5	Naphthalene-mono- and -di-sulfonic acids	3.5	50

Table 3.18. Thermodynamic Functions for the Exchange of Adenosine and Hydrogen on Specific Ionites

Ionite	K_{sw}	$\Delta\Phi°$	$\Delta H°$	$T\Delta S°$
			kJ/mol	
SNK–D	9.1	−11.1	−16.7	−5.6
SNK–D	4.9	−10.8	−18.0	−7.2
SNK–D	4.2	−10.5	−18.8	−8.3
SNK–D	3.2	−10.3	−19.7	−9.4
SNK–E	8.2	−11.4	−13.8	−2.4
SNK–E	4.9	−11.1	−14.7	−3.6
SNK–E	3.9	−10.8	−16.4	−5.6
SNK–E	2.9	−9.0	−17.2	−8.2
SNK–H	6.6	−11.0	−13.4	−2.4
SNK–H	4.6	−10.2	−16.1	−5.9
SNK–H	3.4	−9.5	−17.5	−9.0
SNK–H	2.9	−8.3	−17.6	−9.3

Table 3.19. Sorption Selectivities for Nucleotides (pH $\sim$ pK_1)

Nucleotide	Ionite	K_s ($\bar{N}_1 < 0.1$)
Adenosine-5'-phosphate	KU–2	1.6
	KU–5A	15.4
Uridine-5'-phosphate	KU–2	3.56
	KU–5A	22.3
Thymidine-5'-phosphate	KU–2	0.56
	KU–5A	36.8

doubly charged anions for pH > 6. Studies[299,300] using cationites and anionites have shown that the selectivity of anion exchange is small for the standard anionites both with respect to small ions (chloride) and with respect to different nucleotides. Nucleotides are sorbed selectively in competition with hydrogen ions by polycondensed cationites that contain sulfonated naphthalene residues (Table 3.19). Note in the case of nucleotides that the groupings responsible for the specific interactions with the ionite lie close to the ionogenic groups, and that additional bonds form in between individual resinates. This favors cation exchange, though complicating the determination of the structure of an ionite that sorbs the nucleotide selectively, both differentially and with respect to small ions.

Many alkaloids are comparatively strong bases and may be sorbed by cationites from aqueous solutions and extracts. This, and their differing structures and electrochemical properties, enables both cationites containing sulfo and carboxylic groups and anionites to be used to sorb alkaloids to separate and purify them[301–312]. Quantitative investigations of the region around the sorption maxima of alkaloids on sulfocationites[306,308] have shown that the selectivities for morphine and codeine are good for polycondensed cationities, given that the mole fraction of the organic counter-ion in the ionite is small. However, the selectivity falls as the ionite is loaded up with the alkaloid.

Significant work has been done on selectivity for sulfonamides and various other relatively low-molecular-weight pharmaceuticals using thermodynamics and for a variety of cationites and anionites[313–321].

4

Equilibrium Dynamics of Ion Sorption and Standard Quasi-Equilibrium Frontal Chromatography

Sorption in columns is said to be dynamic as opposed to static, the latter occurring in vessels in which the sorbents and solution are agitated. The difference between the two is that in dynamic sorption the equilibrium is established with the initial solution in most of the column (above the sharp front), while in static sorption the solution is gradually depleted of sorbate and the equilibrium is finally established with a solution that has a different composition than the one originally added to the vessel. A natural result is that the static sorption capacity is lower than the dynamic one, and not just for molecular adsorption, but for ion exchange too.

Let us consider the exchange of two ions with the same valence. We have the ion-exchange equation

$$\frac{\bar{C}_1}{\bar{C}_2} = K_s f(\gamma) \frac{C_1}{C_2} \tag{4.1}$$

and an equation reflecting the invariance of the exchange capacity (at a given solution pH)

$$\bar{C}_1 + \bar{C}_2 = \bar{C}, \tag{4.2}$$

where $\bar{C}$ may correspond to the sorption capacity of the ionite. We also have the equation for the equivalence of the ion exchange, i.e.,

$$C_1 + C_2 = C^0, \tag{4.3}$$

where C^0 is the concentration of the exchanging ions in the original solution.

By comparing Eqs. (4.1), (4.2), and (4.3) we get

$$\frac{\bar{C}_1}{\bar{C} - \bar{C}_1} = K_s f(\gamma) \frac{C_1}{C^0 - C_1} \tag{4.4}$$

or

$$\bar{C}_1 = \frac{\bar{C} K_s f(\gamma) C_1}{C^0 + [K_s f(\gamma) - 1]C_1} \tag{4.5}$$

and two analogous equations for the second component.

Equation (4.5) is very convenient for analyzing the dynamics of ion exchange and can be called the isotherm of dynamic ion-exchange. It shows that there is an unambiguous relationship between the concentration of sorbed ions, $\bar{C}_1$, and the concentration of the ion in the equilibrium solution, C_1, given that the ion exchange is equivalent, that the ionite's exchange capacity is constant, and that the initial solution's concentration is C^0. This relationship can be given in the form of an isotherm with saturation (Figure 1.2). In a static experiment with a low residual concentration C_1 the sorption capacity ($\bar{C}_{1stat}$) is smaller than the sorption capacity in a dynamic regime ($\bar{C}_{1dyn}$) for the same concentration of sorbate ions in the initial solution, C^0.

It should be noted that dynamic sorption is better than static sorption not only from the point of view of sorption capacity, but also because dynamic regimes can be used (with the aid of new techniques) to get specific sorption and desorption by creating sharp boundaries either for one or for a few particular components. Thus columns, if long enough, can be used to separate substances that can form sharp fronts from those that cannot.

4.1. SHARP FRONT FORMATION FOR THE EXCHANGE OF IONS WITH THE SAME VALENCE

The equilibrium dynamics of ion exchange is studied using mass balances for all the exchanging components (4.6) and the ion-exchange isotherm (4.7), viz.,

$$V \frac{\partial C_i}{\partial x} + \varepsilon \frac{\partial C_i}{\partial t} + \frac{\partial \bar{C}_i}{\partial t} = 0, \tag{4.6}$$

where x is the distance from the outermost layer of sorbent in the column,
  ε is the relative volume between the sorbent beads and filled by solvent,
  $\bar{C}_i$ is the concentration of sorbed ion i per unit volume of sorbent,
  C_i is the concentration of ion i in the solution in equilibrium with the ionite,
  V is the flow rate of the liquid in the column, and
  t is time.

$$\bar{C}_i = f(\bar{C}_1, \bar{C}_2, \ldots, \bar{C}_j), \tag{4.7}$$

where $\bar{C}_1, \bar{C}_2, \ldots, \bar{C}_j$ are the concentrations of all the exchanging sorbed ions other than i.

The complexity of studying multicomponent ion-exchange dynamics may be considerably simplified by considering a binary exchange[322]. We shall initially look at the exchange of ions with the same valence. Given that the ionite's exchange capacity is constant and equivalent binary exchange, the isotherm can be given as simply a relationship between the concentration of the sorbed ions of a given type and the concentration of the ion in the solution, Equation (4.5). Thus the mass-balance equations can each be solved independently, and there is no need to solve them together as a system. Hence the problem reduces to an analysis of the mass balances and isotherms of either the ions entering the column (1-ions) or the desorbing ions (2-ions), and this can be compared to the analysis of the sorption dynamics of a single-component system[323-325].

By taking the equations

$$V \frac{\partial C_i}{\partial x} = \varepsilon \frac{\partial C_i}{\partial t} + \frac{\partial \bar{C}_i}{\partial t} = 0$$

and

$$\bar{C}_i = f(C_i)$$

$$\left. \right\} \qquad (4.8)$$

together with the obvious relationships

$$\frac{\partial \bar{C}_i}{\partial t} = \frac{\partial \bar{C}_i}{\partial C_i} \cdot \frac{\partial C_i}{\partial t}$$

and

$$\frac{\partial C_i}{\partial x} = - \frac{\partial C_i}{\partial t} \cdot \left(\frac{\partial t}{\partial x} \right)$$

$$\left. \right\} \qquad (4.9)$$

we can transform the mass balance to

$$\left(\frac{\partial x}{\partial t} \right)_{C_i} = \frac{V}{\varepsilon + \partial \bar{C}_i / \partial C_i} . \qquad (4.10)$$

The equation shows that sharp zones will form if $\frac{\partial^2 \bar{C}_i}{\partial C_i^2} < 0$ for the ions entering the column and $\frac{\partial^2 \bar{C}_i}{\partial C_i^2} > 0$ for the ions leaving the column. The corresponding isotherm is given by Equation (4.5). If the bend in the isotherm does not change (it only becomes more convex or more concave to the concentration axis), then only the initial section as $C_i \to 0$ need be traced, i.e.,

$$\bar{C}_i \Big|_{C_i \to 0} \cong \frac{\bar{C} K_s f(\gamma) C_i}{C_i^0} . \qquad (4.11)$$

A consideration of the form of these isotherms leads us to the conclusion that the $\bar{C}_i$-C_i isotherm is convex to the concentration axis for $K_s f(\gamma) > 1$. The isotherm is concave to the concentration axis for $K_s f(\gamma) < 1$. Bearing in mind Equation (4.10), we may conclude that if we consider the ion entering the column (the displacer ion) to be the 1-ion, then

$$K_s f(\gamma) > 1 \qquad (4.12)$$

must hold for sharp boundaries to form. Since organic ions are usually selectively sorbed, they will, as a rule, form sharp frontal zones. For the desorption of an organic ion, if the replacing ion is a mineral ion and the 1-ion, then (4.12) should be fulfilled. If the 1-ion is the organic ion, then

$$K_s f(\gamma) < 1 \qquad (4.13)$$

should be fulfilled. These criteria are difficult to meet if the ion exchange involves strong electrolytes. A possibility is to use a weak electrolyte to desorb the ion. The isotherm for the exchange of two ions with the same valence will then take the form

$$\frac{\bar{C}_1}{\bar{C}_2} = K_s f(\gamma) \frac{\alpha_1 C_1}{\alpha_2 C_2} ,$$

(4.14)

where α_1 and α_2 are the ionization degrees of the electrolytes including the exchanging ion, and
C_1 and C_2 are the electrolyte concentrations.

Combining (4.14) with (4.2) and (4.3) yields the isotherm of dynamic ion-exchange, which gives the relationship between the component concentrations in the column, in the form

$$\bar{C}_1 = \frac{\bar{C} K_s f(\gamma) (\alpha_1/\alpha_2) C_1}{C^0 + [K_s f(\gamma) - 1]C_1} .$$

(4.15)

If we use this equation to analyze the deformation of the front boundary of an ion zone together with Equation (4.10), we can get a criterion for the formation of a sharp boundary for the desorbing (first) component in the form

$$K_s f(\gamma) (\alpha_1/\alpha_2) > 1.$$

(4.16)

The best and most used method of displacing organic ions from ionites in a column is to lower the level of dissociation of the electrolyte containing the displacing ion (α_2). It is then not difficult to satisfy (4.16) by changing the pH of the solution by using a strong electrolyte as the displacing agent (displacer).

Convex isotherms, or more accurately their limiting (rectangular) form, occur for desorbing ions when cations are being displaced from carboxylic cationites by acids, or when anions are being displaced from weak anionites by alkalis. The methods of frontal displacement corresponding to these two ways of forming sharp boundaries are widely used for laboratory and industrial purposes.

4.2. FORMATION OF SHARP FRONTS DURING THE EXCHANGE OF IONS WITH DIFFERENT VALENCES

In order to consider the dynamics of this situation we must find the form of the dynamic isotherm after combining the ion-exchange isotherm, the equation for the invariance of exchange capacity, and the equation for equivalence of the ion exchange. Such a combination leads to the relationship

$$\frac{c_1^{1/z_1}}{(\bar{c} - \bar{c}_1)^{1/z_2}} = K_s f(\gamma) \frac{c_1^{1/z_1}}{(c^0 - c_1)^{1/z_2}} .$$

(4.17)

This equation describes the behavior of an ion-exchange system involving two strong electrolytes, given that the initial concentration of the 1-electrolyte, which is fed into the tower, is c^0. The plot of $\bar{C}_1$ versus C_1 may be drawn as isotherms similar to those in Figure 1.2. If we limit our analysis to a system whose isotherms have no kinks (i.e., $\frac{\partial^2 \bar{C}_1}{\partial C_1^2} > 0$ or $\frac{\partial^2 \bar{C}_1}{\partial C_1^2} < 0$), and this class includes most ionite/organic counter-ion systems, then we only need to consider the shape of the initial section (with respect to the concentration axis) in order to establish the form of the isotherm, i.e.,

$$\frac{m_1}{\bar{C}_1} \Big|_{C_1 \to 0} = [K_s f(\gamma)]^{z_1}\left(\frac{\bar{C}}{C^0}\right)^{z_1/z_2}. \qquad (4.18)$$

The $\bar{C}_1$-C_1 isotherm must lie above this line and must therefore be convex to the concentration axis if

$$[K_s f(\gamma)]^{z_1}\left(\frac{\bar{C}}{C^0}\right)^{z_1/z_2} > \frac{\bar{C}}{C^0}. \qquad (4.19)$$

This corresponds to the criteria

$$\left.\begin{array}{l} C^0 < \bar{C}[K_s f(\gamma)]^{z_1 z_2/(z_1 - z_2)} \quad \text{for } z_1 > z_2, \\[2em] C^0 > \bar{C}[K_s f(\gamma)]^{z_1 z_2/(z_1 - z_2)} \quad \text{for } z_1 < z_2. \end{array}\right\} \qquad (4.20)$$

If we introduce the concept of a critical concentration, i.e.,

$$C_{cr} = \bar{C}[K_s f(\gamma)]^{z_1 z_2/(z_1 - z_2)}, \qquad (4.21)$$

then the criteria in (4.20) become

$$\left.\begin{array}{l} C^0 < C_{cr} \quad \text{for } z_1 > z_2, \\[2em] C^0 > C_{cr} \quad \text{for } z_1 < z_2. \end{array}\right\} \qquad (4.22)$$

The criteria in (4.20) or (4.22) together with Equation (4.10) define the conditions in which sharp boundaries are formed for the dynamic sorption of the 1-ion and the desorption of the 2-ion because the isotherm corresponding to the incoming component is convex to the concentration axis.

The condition $C_1^0 < C_{cr}$ must be satisfied for organic polyvalent ions to be sorbed in columns such that sharp boundaries are formed when ions with differing valences are being exchanged. Since the solutions, extracts, and culture liquors that are used during purifications and separations are relatively dilute (C_1^0 is small), the sorbate does not overshoot the column until sharp zones appear and the concentration at the outlet of the column grows quickly to the original concentration. Significant difficulties often arise during the desorption of a polyvalent organic ion from a column because the sharp zones are only formed when $C_1^0 > C_{cr}$. For example, the critical concentration for the streptomycin-sodium system (the exchange of a trivalent and monovalent ion) is between 0.2 and 6 N, depending on the Permutit[326]. Hence an ionite must be chosen for which the critical concentration is low. The ions may also be displaced by lowering the ion-exchange constant of the ionite using displacing solutions in organic solvents, or by lowering the ionization of the desorbate electrolyte, or the ionite itself. We considered these in the last section.

Large-scale preparations are only possible if the criteria for sharp-zone formation - Equations (4.12), (4.16), and (4.22) - are satisfied. Figure 4.1 shows how scaling up a column in height increases the concentration of the substance in the eluate. In contrast, when the sharp-zone criteria are not satisfied, the concentration of electrolyte eluted from the column is reduced as the column is scaled up (Figure 4.2). Tests using short columns sometimes yield data that are comparable irrespective of

112

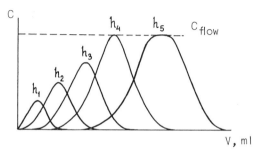

Fig. 4.1. Scheme of how scaling affects frontal desorption in a column.
C_{flow} is the concentration of counter-ion in the solution flow.
Column heights $h_5 > h_4 > h_3 > h_2 > h_1$.

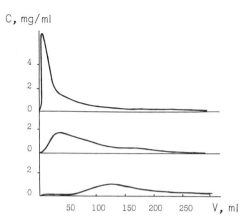

Fig. 4.2. Elution curves for desorption of chlortetracycline from the
SBS-2 ionite in a column. Eluant 1 N HCl in methanol. Height
of ionite layer in column: 1) 1.7 cm; 2) 3.4 cm; 3) 8.7 cm.

whether the zones are diffuse or sharp. However, scaling up the column
distinguishes the regimes markedly and only the sharp-boundary ones are of
any interest for either laboratory or industrial purposes.

It is thus important to be able to find the conditions for sharp
boundaries in industrial scale equipment. The laws we described in this
chapter are satisfied experimentally only for slow flow rates (or on small
ionite beads) since we have only considered <u>equilibrium</u> ion-exchange
dynamics. However, equilibrium can be approximated well in many realizable
and usual experimental conditions.

4.3. FORMATION OF SHARP FRONTS WHEN TWO LIQUIDS ARE INVOLVED

A three-phase system with a mobile phase consisting of two immiscible
liquids can be used to displace the ion-exchange equilibrium further in the
direction of desorption. A rational choice of solvents and conditions can
both displace the equilibrium of the ionite-water system towards desorption
and enhance the displacement by shifting the equilibria in the two co-
existing liquid phases in the direction of transfer from the aqueous phase
to the organic one. Clearly, this must be implemented using a noncoalesc-
ing emulsion in contact with the ionite. The equations describing the
deformation of the ion zones must now allow for two mobile phases[327].

The material balance for the sorption of a one-component substance must in this case be

$$\frac{V}{\delta + \delta'} (\delta \frac{\partial C}{\partial x} + \delta' \frac{\partial C'}{\partial x}) + \delta \frac{\partial C}{\partial t} + \delta' \frac{\partial C'}{\partial t} + \frac{\partial \bar{C}}{\partial t} = 0, \qquad (4.23)$$

where C is the concentration in the aqueous phase,
   C' is the concentration in the organic solvent,
   $\bar{C}$ is the concentration in the ionite phase,
   V is the feedrate of the liquid phase to the column,
   x is the distance from the outermost sorbent layer in the column,
   t is the time,
   δ is the volume fraction of the aqueous phase in the column, and
   δ' is the volume fraction of the organic phase in the column.

For the sorption dynamics of this relatively complicated system to be studied completely two material balances like (4.23) must be solved simultaneously for the two exchanging ions. However, as was the case for the two-phase system[322], we can consider each of the ions (the displacing and displaced) being exchanged in the column independently.

Consider a monovalent/monovalent exchange, for which by analogy with Equations (4.21)-(4.23) the ion-exchange isotherm is valid, the exchange capacity is constant, and the ion exchange is equivalent, i.e.,

$$\frac{\bar{C}_1}{\bar{C}_2} = K_s f(\gamma) \frac{C_1}{\alpha_2 C_2}, \qquad (4.24)$$

$$\bar{C}_1 = \bar{C} - \bar{C}_2, \qquad (4.25)$$

$$\delta C^0 = \delta C_1 + \delta C_2 + \delta' C_2. \qquad (4.26)$$

It is assumed that only the second (desorbing) component is in the two liquid phases.

Combining (4.24) and (4.25) yields a relationship between $\bar{C}_2$ and C_2 only. By using the isotherm equation we get the distribution of the second component between the two liquid phases as

$$\frac{C_2'}{(1 - \alpha_2) C_2} = K_p' = \frac{K_p}{1 - \alpha_2}, \qquad (4.27)$$

and thus the dynamic isotherm can be obtained from Equations (4.24)-(4.27), i.e.,

$$\bar{C}_2 = \frac{\bar{C} \alpha_2 K_s f(\gamma) C_2}{C^0 + [\alpha_2 K_s f(\gamma) - 1 - (\delta'/\delta) K_p' (1 - \alpha_2)] C_2}. \qquad (4.28)$$

In view of this isotherm for dynamic sorption, the material balance of (4.23) can be solved independently for each component. Equations (4.23) and (4.28) for the second (desorbing) component can be transformed to

$$-V' \frac{\partial C_2}{\partial x} = \Delta \frac{\partial C_2}{\partial t} + \frac{\partial \bar{C}_2}{\partial t},$$

where

$$V' = V \frac{\delta + \delta' K_p' (1 - \alpha_2)}{\delta + \delta'}, \qquad (4.29)$$

$$\Delta = \delta + \delta' K_p' (1 - \alpha_2).$$

This equation can be solved for the $\bar{C}_2$–C_2 relationship and the deformation of the sharp zones in a three-phase system defined as

$$\left(\frac{\partial x}{\partial t}\right)_{C_2} = \frac{V'}{\Delta + d\bar{C}_2/dC_2}. \qquad (4.30)$$

The zone boundaries are sharp if for the component being desorbed the $\bar{C}_2$–C_2 isotherm is convex to the concentration axis, i.e., if

$$\frac{d^2\bar{C}_2}{dC_2^2} > 0. \qquad (4.31)$$

The second derivative of (4.28), given that $f(\gamma)$ is constant, takes the form

$$\frac{d^2\bar{C}_2}{dC_2^2} = \frac{2\bar{C}\alpha_2 K_s f(\gamma)[\alpha_2 K_s f(\gamma) - 1 - (\delta'/\delta)K_p'(1 - \alpha_2)]}{C^0[\alpha_2 K_s f(\gamma) - 1 - (\delta'/\delta)K_p'(1 - \alpha_2)]C_2}. \qquad (4.32)$$

Bearing in mind that the denominator cannot be negative, a criterion for the formation of sharp boundaries in a three-phase system must include the degree of dissociation of the desorbate electrolyte, α_2, the partition coefficient, K_p', and the ratio of the volumes of the organic to aqueous phases, δ'/δ. Using Equations (4.30)–(4.32), we obtain the condition that must be satisfied for sharp boundaries to be formed, viz.,

$$\alpha_2 K_s f(\gamma) - (\delta'/\delta)K_p(1 - \alpha_2) < 1. \qquad (4.33)$$

Remember that the "less than" sign in (4.33) arises because we considered the situation for the desorbate. Bearing in mind Equation (4.27) and introducing the notation $\bar{K} = \alpha_2 K_s f(\gamma)$, (4.33) can be transformed to

$$\bar{K} - (\delta'/\delta)K_p < 1. \qquad (4.34)$$

In this way the use of a two-phase desorbing solution enables us to find criteria for sharp boundaries in addition to the one for a two-phase system, by introducing the partition coefficient K_p (or K_p') and the ratio of the two liquid phase volumes.

4.4. SORPTION-DISPLACEMENT FOR A MULTICOMPONENT EXCHANGE IN A COLUMN

If a number of counter-ions are competing for the active centers, the components begin to displace each other. Given that the final partition equilibrium in the column is defined by the selectivity coefficients, the motion of the substance zones will depend directly on the conditions for the formation of sharp boundaries. The best way of seeing this is in a three-component ion-exchange system, with two exchanging ions that have differing valences and are fed into the column in the liquid phase[328]. According to the theory we have presented, the formation of sharp boundaries [cf. Equation (4.22)] is dependent on the selectivity coefficients, the ion charge, and the concentrations of the counter-ions in the feed. As predicted by the theory (Figure 4.3), when a two-component solution of calcium (Ca^+) and trimethylbenzylammonium ($TMBA^+$) cations is fed into a column containing a sulfocationite with sodium counter-ions, the calcium ions will be displaced close to the front of the sodium desorption if $C^0 > C_{cr}$, while the TMBA ions will be displaced if $C^0 < C_{cr}$, sharp boundaries being formed in both cases. The displaced ion will leave the column with a concentration higher than in the feed. This situation is used to separate one component (occasionally two) from a mixture fed into a column[11,329, 330]. Figure 4.4 shows the elution curves of the separation of vitamin B_{12}

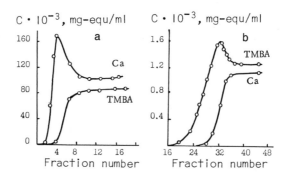

Fig. 4.3. Order of the components in frontal ion-exchange versus total concentration in the initial solution for the trimethylbenzyl-ammonium (TMBA)-calcium system. a) $C^0 > C_{cr}$; b) $C^0 < C_{cr}$.

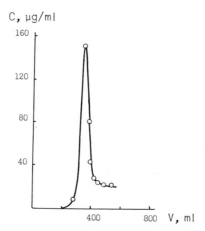

Fig. 4.4. Frontal release process. Sorption of vitamin B_{12} on hydrogen-form SDV-3 sulfocationite.

from a column with hydrogen-form sulfocationite and with the vitamin
extract being filtered from the biomass of propionic acid bacteria by the
column. The concentration of the vitamin in the frontal sorption zone is
8-10 times that of the concentration in the feed solution.

4.5. STANDARD QUASI-EQUILIBRIUM FRONTAL CHROMATOGRAPHY ON IONITES

The specific sorption of a substance from a solution with a column
completely saturated by it together with a specific desorption is basic to
preparative separation and purification, particularly those of complicated
physiologically active substances using permeable ionites. The standard
process for both laboratory and industrial columns involves ionite beads
0.1-0.5 mm in radius (laboratory work) or 0.25-2 mm in radius (industrial
scale). To get the quasi-equilibrium regimes in which the zone boundaries
are described by the equations we have derived, the flow rate of the
solution through the column must be quite low. During the sorption of
physiologically active substances from extracts, culture liquors, or other
dilute solutions when the ionite bead diffusivities for the ions are
between 10^{-8} and $10^{-9} cm^2/s$, the flow rate is between 100 and 500 ml per
square centimeter of column cross-section per hour. At this flow rate the
boundaries of the ion zones migrate at steady-state and with relatively
little spreading if the criteria of (4.12), (4.16), (4.22), or (4.34) are
satisfied. In industrial installations, in which the bead sizes are
larger, the saturation of the sorbents with the substances being separated
is implemented using a series of columns, the ions leaving one column in
front of a relatively spread zone to be sorbed in the next. Only the
components in a completely saturated column whose sorbent is in equilibrium
with the feed solution are desorbed.

Sharp zones can be formed with saturation if very specific sorbents
for the absorbate are used. Obviously the process should be intensified
and accelerated, and we shall consider this in the next chapter from the
point of view of kinetics and dynamics. We shall also consider in the next
chapter the complications that can arise due to the low diffusivities of
the ions in the ionites.

The problem of selective desortion is crucial to a preparative process. The very use of a specific ionite to achieve selective sorption
hampers the establishment of the conditions needed for desorption with
sharp boundaries. Thus, given a monovalent/monovalent exchange of strong
electrolytes, for which an ionite has been obtained that has a selectivity
for the counter-ion being sorbed which is tens, hundreds, or even thousands
of times that of the original counter-ion on the ionite, a system must be
found, in order to desorb the ion with sharp boundaries, in which the
selectivity is reduced to less than 1. The more specific the ionite used
for the sorption, the more difficult it is to find a system to desorb the
ion with sharp boundaries.

There is also a problem as regards the solution flow rate. During the
desorption the desorbate concentration rises about that of the original
solution being sorbed; thus the desorbate zone may be comparable in width
to the boundary spreading. Thus the spread region during the desorption
must not be large, and this is achieved in the standard process by using
solution volumetric flow rates of 25-30 $ml/cm^2 \cdot h$.

The choice of desorption conditions to get sharp boundaries is crucial
for achieving the fullest desorption and highest concentration of the
substance in the eluate. Figure 4.5 shows the elution curves for the
desorption of oxytetracycline for columns of different heights. Sharp-
boundary conditions are made to prevail between the ammonia buffer and the

antibiotic and the same concentration of antibiotic is fed into the columns[33]. The curves demonstrate that the desorption zone moves along the column without spreading and without a chromatograph tail. The reverse situation is shown in Figure 4.2, where the prevailing conditions permit spreading[326]. The distortion of the boundaries, i.e., spreading, gets worse as the column height is increased.

Scaling up is even more favorable to desorption when sharp-boundary conditions prevail and the column is completely saturated with the substance being separated (Figure 4.6). This is because the concentration of the substance being separated increases as the tower height is increased, while the lower concentration fraction is reduced. Thus the yield of the main fraction is increased, the fraction being further treated by precipitation, crystallization, and lyophilization. Figure 4.1 shows schematically the effect of scaling on frontal desorption. It can be seen that the limiting concentration in the eluate of the desorbate is the concentration of the displacer ion in the feed. The formation of sharp boundaries is important, indeed vital, when scaling up preparative processes.

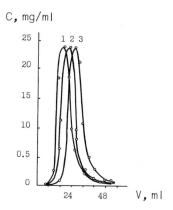

Fig. 4.5. Elution curves for desorption of oxytetracycline. Eluant 0.01 N ammonium borate (pH 9). Desorption rate 50 ml/h·cm². Height of ionite layer in column: 1) 5.7; 2) 8.1; 3) 11.8 cm.

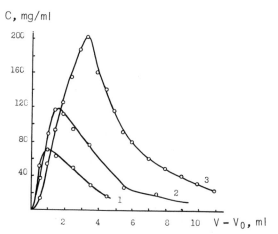

Fig. 4.6. Elution curves for kanamycin desorption from a column with KB-4P-2 carboxylic cationite. Eluant 2 N ammonia. Desorption rate 25 ml/h·cm². Height of ionite layer in column: 1) 1.6; 2) 3.2; 3) 8.0 cm.

If the desorption of a substance in a column can occur in two fairly similar regimes, one characterized by sharp boundaries, the other by spread boundaries, then on a small (laboratory) scale using a short tower (several centimeters) the two regimes will not be very different from the point of view of the eluate curves, the maximum concentration of substance in the eluate, or the yield. Scaling up the process to use a column several tens of centimeters high changes the situation dramatically. The sharp-boundary regime has better parameters all around with higher concentrations and yields, whereas the spread-boundary regime has much poorer parameters; indeed these parameters have, as a rule, lost all practical meaning. Further scaling increases the division, making it impossible to use a regime for dynamic sorption in which sharp boundaries are not formed.

The experimental work on frontal desorption involving physiologically active substances has led to the development of several methods for obtaining active eluates together with the complete desorption of the substance being separated [115,254,257,289,293,326,328,330,332-339].

To prepare physiologically active substances using the easily implemented selective sharp-boundary sorption, the usually difficult problem of getting sharp-boundary desorption is often overcome by altering the ionization of the desorbate ions. During cation exchange this is done by increasing the pH of the eluate beyond the threshold given by the dynamic criteria, while for anion exchange the pH is reduced to the requisite levels. Lowering the pH is the most convenient and accepted method for weak cationites (adding acids for the carboxylic cationites), while raising the pH is best for the weak anionites.

The simplest criterion (4.13) and its inverse are exceedingly difficult to implement for frontal sorption and desorption if very specific ionite/counter-ion systems are used for which the ion-exchange constants are large. However, in individual cases it is sometimes possible. Thus for the novobiocin-chloride system changing from an aqueous to a water-alcohol solution reduces the ion-exchange constant a thousandfold, and with an 85-90% concentration of ethanol the constant becomes less than unity (Figure 4.7). This makes it possible to have sorption with a selectivity greater than a 1000 and then get a full displacement of novobiocin (Figure 4.8) without a chromatograph tail. This process can be scaled up.

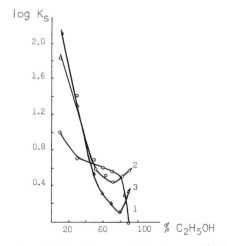

Fig. 4.7. Sorption selectivity for novobiocin on 1) FAF, 2) AV-17, and 3) AV-16 versus concentration of ethyl alcohol in solution.

Sequential desorption of simultaneously sorbed components is more complicated, especially when each stage of the displacement must lead to the formation of sharp boundaries for the ion zone of only one component while the conditions spread the boundaries for the other components, thus stopping them from markedly migrating down the column. If a sorption is run involving such a system, and with a certain undersaturation of the column, thus leaving a free zone at the bottom of the column to collect slowly (with spread boundaries) moving components, then two, three, or four (sometimes) closely related components (in terms of their properties) can be completely separated using a preparative process with a very high absorptivity for the substances being separated. For example, adenosine mono-, di-, and triphosphate have been displaced in sequence from an amphoteric ionite in this way (Figure 4.9).

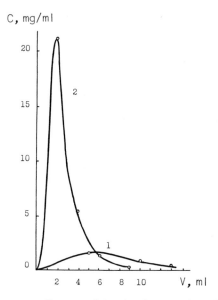

Fig. 4.8.   Elution curves for novobiocin desorption from FAF.   Eluants:
1) 2.5 N NaCl and 2) 5.0 N NaCl in 90% ethanol.

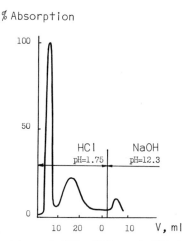

Fig. 4.9.   Separation of ATP, ADP, and AMP in a column with KU-5 cationite.

Heteronet carboxylic cationites (see Chapter 2) in the form of relatively large beads can be used to prepare peptides and proteins at standard solution flow rates, i.e., using standard quasi-equilibrium chromatography. Since the diffusivities of proteins in permeable ionites is usually less than 10^{-9} cm^2/s, both sorption and desorption are carried out with slower flow rates than they are for smaller organic ions, whose diffusivities in gels are usually around 10^{-8} cm^2/s. Some examples of the preparation of proteins (enzymes) and polypeptides are the separation of acidic and neutral proteases from the culture liquor of *Bac. subtilis* (Figure 4.10), neuraminidase (Figure 4.11), urokinase direct from urine (Figure 4.12), and the peptide thymarin from thymus extract (Figure 4.13). These were obtained to a very high degree of purity suitable for medical purposes in one or two stages and directly from culture liquors and extracts. In each case the principle of changing the ionization level of the cation form of the protein, or even adding or reversing the charge on the

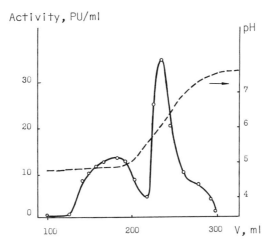

Fig. 4.10.   Separation of acidic and neutral proteases on Biocarb-T. Stepwise elution by a phosphate buffer solution, pH 4.9-7.5.

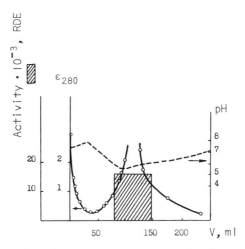

Fig. 4.11.   Separation of neuraminidase from extracts in a column with Biocarb-T.   Desorption by 0.1 M phosphate buffer solution, pH 7.4.   RDE is a unit of the activity in erythrocyte destruction reactions.

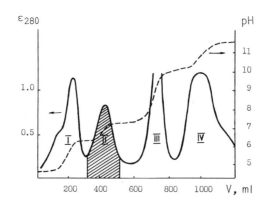

Fig. 4.12. Separation of urokinase in a column with Biocarb-T. Desorption by phosphate buffer solution. Regions: urokinase II, and protein admixtures I, III, and IV.

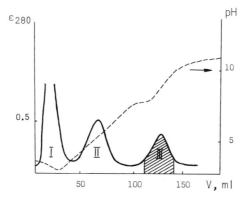

Fig. 4.13. Separation of immunostimulater (III) from thymus extract in a column with Biocarb-T. Eluant gradient between pH 2.5 and pH 9.5. I and II for protein admixtures.

Table 4.1. Diffusivities of Organic Ions in Ionites

Organic ion	Ionite	Sorption		Desorption	
		pH	D, cm^2/s	pH	D, cm^2/s
Oxytetracycline	Dowex-50	2.5	$0.64 \cdot 10^{-9}$	11	$0.96 \cdot 10^{-7}$
	SBS	2.5	$7.5 \cdot 10^{-9}$	11	$2.0 \cdot 10^{-7}$
Ristomycin	CPA	7	$2.8 \cdot 10^{-9}$	11	$5.4 \cdot 10^{-7}$
	SBS-3	7	$1.6 \cdot 10^{-9}$	11	$2.4 \cdot 10^{-7}$
	SDV-3	7	$2.5 \cdot 10^{-9}$	11	$2.8 \cdot 10^{-7}$
Hygromycin	KB-4P-2	7	$2.5 \cdot 10^{-8}$	11	$2.9 \cdot 10^{-7}$
	Wofatite SP-300	7	$3.0 \cdot 10^{-8}$	11	$5.4 \cdot 10^{-7}$
	SDV-3T	7	$4.4 \cdot 10^{-8}$	11	$7.7 \cdot 10^{-7}$
	KU-2	7	$3.6 \cdot 10^{-8}$	11	$5.3 \cdot 10^{-7}$

co-ion, was used. The latter was important because the sign of the charge is vital for the kinetics of the ion exchange. Table 4.1 demonstrates that changing the charge of the dipole ions of oxytetracycline and of the organic bases of ristomycin and hygromycin B from being counter-ions to being co-ions (or to neutral molecules) leads to an increase in the rate of diffusion of the ions in ionites[340]. This enables the system to approach the quasi-equilibrium regime which is described by the equilibrium theory of ion-exchange dynamics.

5
Kinetics and Non-Equilibrium Ion-Exchange Dynamics

5.1. THE DIFFUSION OF ORGANIC IONS IN IONITES

We demonstrated in the last chapter that the criteria from the equilibrium theory of sorption dynamics can only be satisfied experimentally for the chromatographic preparation of physiologically active substances (and then only approximately) given very slow solution flow rates (25-200 ml/cm^2·h). However, in these regimes and using beads 0.2-2 nm in diameter the zone boundaries are still significantly spread, even for steady-state zone and boundary migrations. The width of the zone and the extent of the boundary along the column are quite significant. A consequence of this is that the zones of closely related substances overlap.

In the absence of an interphase equilibrium the situation is governed by the kinetics of mass transfer. We turn now to a model of the mass transfer kinetics and how to describe it mathematically.

5.1.1. A Model of the Kinetics of Ion Exchange

Analyzing the kinetics of ion exchange together with information gathered from developing separation and purification methods improves our understanding of the mechanisms of ion exchange and the structure and properties of ionites.

In all complicated systems like ion-exchange ones the number of factors influencing a process is very large. This creates immense difficulties in the corresponding mathematical model and when trying to solve concrete problems. Hence any model that can be used to interpret kinetic and dynamic data involving organic ions must be very much simplified. On the other hand, the experimental research in this field is not at present accurate enough to make a more accurate model preferable.

The most widespread point of view is that the ion-exchange rate is determined by various interphase mass-transfer mechanisms, i.e., the convective diffusion of the sorbate in the solution boundary layer (external diffusion) and the diffusion of the sorbate within the ionite (internal diffusion). However, sometimes the exchange itself (the chemical stage) may make as significant a contribution to the ion-exchange kinetics as the mass transfer processes do[16]. Usually the idea of a limiting stage

is used when working on the kinetics of interphase exchange. This is the
process which determines the time when the equilibrium is established, and
thus the whole problem reduces to a simple special case. The only thing
needed is to find out how to determine which type of limitation a given
system has.

Consider the transport model of interphase exchange kinetics. Since
the diffusion in the boundary layer around a bead and diffusion within the
bead are consecutive stages, the slower of the two processes will determine
the rate of the diffusion as a whole. One of the best ways of determining
experimentally which is the limiting process is the method of broken phase
contact[101], Figure 5.1. If the internal diffusion is limiting, then
after the contact has been broken the concentration within the bead will
flatten out, and so when the contact is re-established the exchange rate
will increase. If the external diffusion is limiting, then there is no
concentration gradient within the bead at any time during the process;
hence a break and resumption of contact will have no effect on the rate of
exchange.

A number of models take into account the effect of electric fields on
ion-exchange diffusion. In these models the transfer of the exchanging
ions in opposite directions is seen as a codiffusion coordinated by the
presence of the electric fields of the charges. The driving force for the
process, besides the concentration gradient of the charged particles, is
the gradient of the electrical potential[341-345]. This theory requires
the fields to have a significant influence on the concentration of the ions
within the ionite and explains the observed asymmetry in the forward and
reverse kinetics of ion exchange. However, a number of papers have shown
that the influence of the fields on effects averaged by integration over
the volume of a sorbent bead (e.g., on the degree of completion versus time
relationship) is not as large as might be expected from the theory, even
for strongly acidic ionites[346-349]. This sort of integral relationship
is used for a variety of purposes such as formulating and solving problems
involving multi-interaction sorption, solving inverse problems, i.e.,
finding process parameters from experimental kinetic curves, and constructing a theory for dynamic sorption. Moreover, a suitable choice of the
diffusivity D_{eff} can make the isotopic kinetic curve roughly converge to
the curve corresponding to the codiffusivity versus concentration relationship. All this taken together allows us in most cases to use a model that
disregards the influence of the electrostatic fields and that is identical
to the model used in the theory of molecular adsorption.

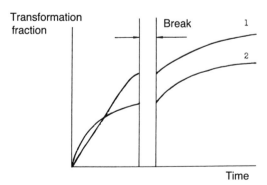

Fig. 5.1.   Experimental investigation of the limiting kinetic stage
            for ion exchange by breaking the contact between the phases.
            Curves 1) film kinetics; 2) gel kinetics.

Most work on the kinetics of ion exchange uses the model of ion diffusion in a quasi-homogeneous medium without sources. This is called the gel diffusion model or diffusion in a solid electrolyte solution. It was first developed for isotopic exchange[350,351], and then extended to exchanges of ions with different properties[352,353].

The mathematical description of the model involves Fick's first and second laws of diffusion:

$$\vec{J} = - D \text{ grad } \bar{C}, \qquad (5.1)$$

$$\frac{\partial \bar{C}}{\partial t} = D\Delta\bar{C} \text{ (for the ionite beads)}, \qquad (5.2)$$

where $\vec{J}$ is diffusion flow density in the sorbent phase,
$\bar{C}$ is the point concentration in the sorbent phase at time t, and
Δ is the Laplace operator.

To describe the mass transfer in a boundary layer (external and mixed diffusion) it is assumed that the thickness of the layer δ is constant and small over the whole surface of the bead, and that equilibrium between the solution and the sorbent surface is established instantaneously. Hence

$$\vec{J} = - \frac{D}{\delta} (C - C*), \quad \vec{J}\big|_{r=R} = \vec{J}\big|_{r=R}, \qquad (5.3)$$

where $\vec{J}\big|_{r=R}$ and $\vec{J}\big|_{r=R}$ are diffusion flow densities, respectively, in the boundary layer and at the surface of a sorbent bead of radius R,
C is the concentration in the bulk of the agitated solution, and
$C*$ is the concentration in the solution at the surface of the bead.

Another system was examined. It has a finite heterogeneous distribution of sorbent within the ionite bead. For example, it is assumed that the sorption occurs in the surface layer of the bead and is absent at its core. This system is called the nucleus-cloud model and is used

1) when the dense core is formed by the very sorption of organic ions[58,113,114,354,355]; and
2) when the nucleus is specially included in the bead as it is made (surface-layer ionites), the idea being to reduce the mean diffusion path and thus enhance the kinetic parameters of the sorbent[18,356,357].

It it necessary in these cases to remember that the boundary condition of zero for the diffusion stream is not at the center of the bead but at the interface between the sorbing and nonsorbing materials:

$$\vec{J}\big|_{r=R-\ell} = 0, \qquad (5.4)$$

where ℓ is the thickness of the layer, and R is the radius of the bead.

Solutions to kinetic problems are usually sought in the form of the degree of completion $F = Q_t/Q_\infty$ as a function of time; Q_t and Q_∞ are, respectively, the quantity of sorbate absorbed at time t and that at equilibrium. If the function is linear, the exact solution for non-steady-state sorption can be written as

$$F = 1 - \sum_{n-1}^{\infty} B_n e^{-\mu_n^2 F_0}, \qquad (5.5)$$

where B_n is a factor that depends on the bead geometry and the root μ_n of the corresponding characteristic equation,
$F_0 = Dt/L^2$ is the generalization of time (Fourier's number), and L is the characteristic dimension of the bead (L = R for a spherical bead with radius R, and L = ℓ for surface-layer sorbents with sorbable thickness ℓ).

Usually the solution is not used in the form of (5.5) for practical purposes; instead asymptotic approximations for long and short periods are used. Over long periods (large Fourier numbers) only the first term of the series is used in the solution, viz.,

$$F \approx 1 - B_1 e^{-\mu_1^2 F_0}. \tag{5.6}$$

This approximation corresponds to what is called the regular kinetics regime. The regular regime is widely used in hydrodynamics and the theory of heat transfer[358-360]. On the other hand, for small Fourier numbers it is possible, using the operator method, to obtain a solution for an initial irregular regime. Regular and irregular regimes can be generalized to dynamic sorption.

In addition, the method of statistical moments has recently been widely applied to the analysis of kinetic and dynamic sorption[361-369]. The advantage of the method is that analytical expressions for the statistical moments can be obtained relatively simply even when the exact solution is unknown. The initial k-th order moment of the kinetic curve is

$$\bar{t}_k = \int_0^\infty t^k \frac{dF}{dt} \, dt. \tag{5.7}$$

The statistical moments can be calculated from experimental kinetic curves by graphical integration; $\bar{t}_k$ is equal to the area bounded by the kinetic curve plotted in $F-t^k$ coordinates, the ordinate axis, and the straight line $F = 1$.

5.1.2. Internal Diffusion Kinetics

Limitations due to the kinetics of internal diffusion most often occur when large organic ions are being sorbed. An analysis of Equation (5.5) in the case of a linear sorption isotherm yields an expression describing the initial section of the kinetic curve (the irregular regime), i.e., F is linearly dependent on $\sqrt{t}$, which it is between F = 0 and F = 0.2-0.3. For spherical beads the expression is

$$F = 6(1 + w) \sqrt{F_0/\pi} - 3(3w + 1)(1 + w)F_0, \tag{5.8}$$

while for a spherical layer it is

$$F = \frac{6(1 + w)}{1 + \rho + \rho^2} \sqrt{F_0/\pi} - \frac{3(3w + 1 - \rho^3)(1 + w)}{(1 + \rho + \rho^2)^2} F_0, \tag{5.9}$$

where $w = \bar{V}\bar{C}_0/VC = \bar{V}K_d/V$ is the limitation of the solution volume,
$\bar{V}$ and V are, respectively, the volumes of the ionite (excluding any non-sorbing parts) and the solution,
$\bar{C}_0$ is the concentration of sorbate in the sorbent in equilibrium with the sorbate in the solution with its original concentration C_0, and
ρ is the relative radius of the non-sorbing part of the bead,
$\rho = 1 - \ell/R$.

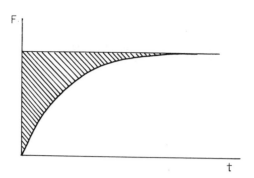

Fig. 5.2.   Graphical calculation of the average sorption time for the
            first statistical moment.

Most experimental investigations have been done for $w \ll 1$, which is true
for large solution volumes and moderate K_d.   In this case it is possible,
given that F is linearly dependent on $\sqrt{t}$, to obtain Equations (2.11) and
(2.14), which are convenient for calculating the diffusivities.

    In order to evaluate the kinetics of a process on average it best to
determine a value of D, averaged over the whole kinetic curve.   This is
quite easy using moments.   For internal diffusion kinetics a knowledge of
the first statistical moment of the kinetic curve - the average sorption
time - is sufficient, i.e.,

$$\bar{t}_1 = \int_0^\infty t \frac{dF}{dt} \, dt = \int_0^1 t dF. \qquad (5.10)$$

This can be found graphically as the area under the kinetic curve in F-t
coordinates (Figure 5.2).   The average value of D may be calculated from
the expression for the average sorption time for internal diffusion, i.e.,
for spherical beads

$$\bar{t}_1 = \frac{R^2}{15(1 + w)D} , \qquad (5.11)$$

and for spherical layers

$$\bar{t}_1 = \frac{(1 + 3\rho + 6\rho^2 + 5\rho^3)}{15(1 + w)(1 + \rho + \rho^2)} \cdot \frac{\ell^2}{D}. \qquad (5.12)$$

5.1.3. Mixed Diffusion Kinetics (Linear Sorption Isotherm)

    The model we described above is kept, but the condition at the phase
boundary is written as

$$D \frac{\partial \bar{c}}{\partial r}\Big|_{r=R} = \frac{D_s}{\delta} [C(R,t) - C_0], \qquad (5.13)$$

where D_s is the diffusivity in the solution, and δ is the thickness of the
diffusion layer in the solution.   An exact solution for this problem mainly
has the form of (5.5).   During the solution we shall use Biot's diffusion
criteria (Bi), which are the ratios of the contributions of the internal
and external diffusions.

For spherical beads we have

$$Bi = \frac{D_s R}{D \delta K_d} \qquad (5.14)$$

and for spherical layers

$$Bi = \frac{D_s \ell}{D \delta K_d} . \qquad (5.15)$$

Given the same accuracy that is permissible in experiments with large organic ions, we may conclude that the kinetics are limited by purely internal diffusion within the bead for $Bi > 20$ and purely external diffusion for $Bi \leq 1$. In the case of $Bi \leq 1$ we have film kinetics, viz.,

$$F = 1 - \exp[- \frac{3D_s}{(1 + \rho + \rho^2) \delta \ell K_d}]. \qquad (5.16)$$

Statistical moments are also useful for the analysis of mixed diffusional ion-exchange. It is then necessary to know analytical expressions for the initial first $(\bar{t}_1)$ and second $(\bar{t}_2)$ order moments of the kinetic curves. The average value of the sorption time is the sum of the average sorption times for purely external and purely internal diffusion, i.e.,

$$\bar{t}_1 = \bar{t}_1^{in} + \bar{t}_1^{ex}. \qquad (5.17)$$

In contrast, the second-order moments are not additive. If the solution volume is limited, we have

$$\bar{t}_{2,w} = \frac{\bar{t}_2}{(1 + w)} - \frac{2w}{(1 + w)^2} (\bar{t}_1)^2, \qquad (5.18)$$

where

$$\bar{t}_2 = \bar{t}_2^{in} + 4\bar{t}_1^{in}\bar{t}_1^{ex} + \bar{t}_2^{ex},$$

$$\bar{t}_2^{in} = \frac{2(2 + 10\rho + 30\rho^2 + 49\rho^3 + 35\rho^4)}{315(1 + \rho + \rho^2)} \cdot \frac{\ell^4}{D^2},$$

$$\bar{t}_2^{ex} = \frac{2(1 + \rho + \rho^2)^2}{9Bi^2} \cdot \frac{\ell^4}{D^2}, \qquad (5.19)$$

$$\bar{t}_1^{in} = \frac{1 + 3\rho + 6\rho^2 + 5\rho^3}{15(1 + \rho + \rho^2)} \cdot \frac{\ell^2}{D},$$

$$\bar{t}_1^{ex} = \frac{1 + \rho + \rho^2}{3Bi} \cdot \frac{\ell^2}{D}.$$

All the above formulas were derived for a linear isotherm, i.e., for unselective sorption. However, selective sorption (with convex isotherms) is much more common. It is possible to get an analytical solution only for very convex (rectangular) isotherms[16].

If the sorption is from a limited volume of solution onto a spherical bead $(\rho = 0)$, the initial part is described by the equation

$$- \frac{1}{w} \ln(1 - wF) = 3BiF_0.$$ (5.20)

The inverse task - finding the kinetic parameters of a process from experimental curves - can be done if the time t*, which is when F stops being linearly dependent on time, can be determined from the curve. Then by simultaneously solving Equation (5.20) for $F_0 = F_0^* = Dt*/R^2$ and the coupling equation

$$\frac{1}{w} + \frac{3\exp(-3wBiF_0^*)}{\sqrt{3wBi} \cot \sqrt{3wBi} - 1} - 2Bi \sum_{n=1}^{\infty} \frac{\exp(-\mu_n^2 F_0^*)}{\mu_n^2 - 3wBi} = 1,$$ (5.21)

where $3wBi \neq \mu_n$, μ_n being a non-zero root of the equation $\mu \cot \mu = 1$, we can find the Bi and F_0^* and then using the known values of F*, t*, R, and w, we can calculate the effective internal diffusivity.

5.1.4. Sorption Kinetics in Ionites with Structural Heterogeneity

Although the model we have presented can strictly be applied to diffusion in gels, which are quasi-homogeneous, it is not suitable for many of the materials that have appeared recently, such as the macronet, macroporous, and biporous (bidispersed) ionites. An account of the morphological heterogeneity of a sorbent in the kinetics of ion exchange is possible in biporous sorbents[370,371]. The diffusion model used in this case assumes the pores are distributed by size in two modes. A sorbent bead is supposed to be a sphere formed from a large number of particles, microbeads, whose dimensions are several-fold smaller than those of the bead itself. The channels between the particles are termed the "transport pores," and the microbeads themselves have pores but with much smaller dimensions. This system has two average sorption times, i.e., $T_i = R_i^2/D_i$ and $\tau_a = R_a^2/D_a$, where R_i is the radius of the bead, R_a is the radius of a microbead, D_i is the diffusivity inside the transport pores, and D_a is the diffusivity in the microbeads.

Again using statistical moments we obtain expressions for the first and second moments of the kinetic curves for sorption from a bounded solution volume into both the transport and microbead pores, assuming a linear isotherm[370]. Thus

$$\bar{t}_1 = \frac{1}{15(1 + w)} (T_i + B\tau_a),$$ (5.22)

$$\bar{t}_2 = \frac{1}{(1 + w)^2} \{\frac{4}{315} (1 + \frac{3}{10} w)T_i^2 + \frac{4}{225} BT_i\tau_a + \frac{4}{315} [1 + (1 - \frac{7}{10} B)w]B\tau_a^2\},$$

where

$$B = [1 + \frac{3(1 + K_1)}{AK_2}]^{-1}$$

and for most industrial sorbents B $\simeq$ 1[372],
A is the number of microbeads per unit volume,
K_1 is Henry's constant for the transport pores, and
K_2 is Henry's constant for the microbeads.

130

These equations can be solved simultaneously in order to find τ_a and T_i. However, an effective diffusivity is needed in order to compare these with other sorbents. A theoretical analysis[373] shows that the average effective diffusivity, i.e.,

$$\frac{R_i^2}{(aR_i^2/D_i) + (bR_a^2/D_a)}, \qquad (5.23)$$

depends on the diffusivities in both the transport pores and the micro-beads. Here a and b are coefficients that depend on the form of the microbeads, and if spherical and for $b \simeq 1$ we have $a \simeq b \simeq 1$.

The creation of an ion-exchange material in which the ionite is dispersed in a granulated, inert, porous matrix entailed a consideration of its sorption kinetics. It is easily shown that a very useful ratio of the dimensions of the microbeads to those of the bead itself can be derived from the diffusivities for the ions in the microbeads and the transport channels using (5.23), i.e.,

$$\frac{aR_i D_a}{bR_a D_i} \geqslant 1. \qquad (5.24)$$

The effective diffusivity in a biporous sorbent will thus differ from the diffusivity in the inert material by less than an order of magnitude. Moreover, in several cases it is possible to estimate $\bar{D}_{eff}$ for biporous sorbents using the formulas for the internal diffusion in gels. Such an estimate may be made if $\tau_a < T_i$, which is the case if the kinetic curve, plotted in F-$\sqrt{t}$ coordinates, is initially linear[374].

5.1.5. Experimental Investigations of the Diffusivities of Organic and Physiologically Active Ions in Ionite Beads

We shall now consider some of the parameters that are important in practical chromatographic sorbents. We shall look at both the relatively small amino acid and antibiotic counter-ions, and at macromolecular protein counter-ions. The influence of crosslink ratio and swelling factor on the diffusion rate of methionine is shown in Table 5.1[396]. These diffusiv-ities are characteristic of the sorption of organic ions. The drop in diffusion rate due to the increased crosslink ratio is typical for gel ionites. Thus for a small DVB concentration in the ionite (2%) the ion exchange is very fast with a diffusivity of around 10^{-7} cm^2/s, which can be used to achieve a quasi-equilibrium dynamic regime in a column, albeit for small solution flow rates, while for 8-16% DVB concentrations the diffusiv-ities fall to 10^{-9}-10^{-10} cm^2/s. At these diffusion rates the kinetic spreading of the ion zones is dominant, and as we shall see it is extremely difficult to attain quasi-equilibrium regimes in a column, even when using small-diameter ionite beads.

The complete opposite is true for the diffusion (quasi-diffusion) of methionine in macroporous ionite beads. For 10-12% of DVB and a limited concentration of porogen (heptane) in the synthesis mixture, a macroporous structure is formed in which organic ions can move both in the channels and in the bulk of the copolymer network surrounding the channels. Since the migration speed in the channels is large and the diffusion path from the channel walls small, the effective quasi-diffusivity, using spheres to model the process as a whole, is moderate, being 10^{-8} cm^2/s. A change to dense nets and rigid channels, coming with a DVB concentration of 40%, alters the ion exchange. Methionine ions (and some other organic ions) diffuse through the transport channels and are only sorbed onto the channel surfaces. As a result, the effective diffusivity achieves the startlingly high value of 6.6×10^{-7} cm^2/s, which is combined with the much lower sorption capacity of this ionite for methionine compared to that of the gel ionites.

Table 5.2 demonstrates that oxytetracycline and insulin diffuse by a different mechanism. The macromolecule of insulin diffuses slowly in a gel cationite and relatively quickly in the transport pores of macroporous sulfocationites, when it is only sorbed onto the macropore surfaces. In contrast, oxytetracycline diffuses faster in gel cationites than in macroporous sulfocationites, which contain 20% DVB and which can sorb it both on the channel surfaces and in the network structure around the channels. The oxytetracycline diffuses very slowly in structures densely knit by large concentrations of crosslinking agents.

Gel ionites made with small concentrations of crosslinking agent (usually 1-3% DVB) are the most widespread for sorbing and preparing organic ions with molecular masses of less than 600-800 daltons. Table 5.3 gives diffusion parameters of these ionites for the tetracyclines, a typical diffusivity being on the order of 10^{-7} cm^2/s. Nucleotides, nucleosides, alkaloids, and many other ions with similar molecular masses have much the same diffusivities.

Macronet ionites, which use long-chain crosslinking agents such as hexamethylenebismethacrylamide (Figure 5.3), have certain advantages with respect to permeability for organic ions of moderate size. The sorption of streptomycin onto these cationites is characterized by diffusivities of around 10^{-7} cm^2/s. Macromolecular proteins often only diffuse into a limited external layer of ionite. Calculating effective diffusivities by

Table 5.1. Diffusivities of Methionine in the Gel Cationite KU-2 and in the Macroporous Cationite KU-23

Ionite	% DVB	% Porogen with respect to liquid phase	K_{sw} (volumetric)	D, cm^2/s
KU-2	2	–	3.4	$1.7 \cdot 10^{-7}$
KU-2	4	–	2.9	$4.0 \cdot 10^{-8}$
KU-2	8	–	2.1	$3.6 \cdot 10^{-9}$
KU-2	16	–	1.7	$1.0 \cdot 10^{-10}$
KU-23	10	60	2.6	$2.0 \cdot 10^{-8}$
KU-23	12	80	3.0	$2.9 \cdot 10^{-8}$
KU-23	15	100	1.6	$6.0 \cdot 10^{-8}$
KU-23	40	100	1.4	$6.6 \cdot 10^{-7}$

Table 5.2. Diffusivities of Oxytetracycline and Insulin in the Gel Sulfocationite SDV-T and the Macroporous KU-23

Counter-ion	Mol.mass	D, cm^2/s	
		SDV-T 4% DVB	KU-23 20% DVB
Insulin	12000	$1.0 \cdot 10^{-11}$	$1.1 \cdot 10^{-9}$
Oxytetracycline	460	$4.8 \cdot 10^{-8}$	$2.6 \cdot 10^{-9}$

Table 5.3. Diffusivities of Tetracyclines in KRS-2 Sulfocationite
(2% DVB)

Counter-ion	D, cm^2/s
Tetracycline	$2.5 \cdot 10^{-8}$
Oxytetracycline	$1.4 \cdot 10^{-8}$
Chlortetracycline	$1.5 \cdot 10^{-8}$
Oxyglucocycline	$0.9 \cdot 10^{-8}$

Table 5.4. Diffusivities of Macromolecules in Highly Permeable Biosorbents

Diffusing ion	Ionite	D, cm^2/s
Terrilytin	Biocarb-T	$0.7 \cdot 10^{-9}$
Lysozyme	Biocarb-T	$6.0 \cdot 10^{-9}$
Blood albumin	Biocarb-T	$0.2 \cdot 10^{-9}$
Blood albumin	Heteroporous anionite	$3.6 \cdot 10^{-9}$
Insulin	Heteroporous anionite	$2.0 \cdot 10^{-9}$
Heparin	Heteroporous anionite	$1.8 \cdot 10^{-9}$

assuming equilibrium loading of spherical ionite beads leads to the experimentally observed relationship between the diffusivities and the bead radius[114]. If it is assumed that only the outer layer, with a thickness depending on the bead radius, is loaded at equilibrium by proteins, then the diffusivities obtained are constant.

It is entirely understandable that even when the highly permeable biosorbents, which we discussed in Chapters 2 and 4, are used the diffusivities of macromolecules are still less than about a tenth of those of smaller organic ions and amount to something on the order of 10^{-9} cm^2/s (Table 5.4). It should therefore be emphasized that both globular proteins and the flexible coils of heparin diffuse in heteronet ionites at comparable speeds.

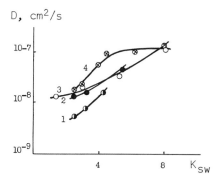

Fig. 5.3.   Influence of crosslink ratio on the rate of diffusion of streptomycin in ionite beads for different carboxylic cationites. Curves 1) KB-4; 2) CPA; 3) KB-2; 4) KMDM-6 (30% acetic acid).

Naturally, in order to realize a macromolecular separation (say proteins and enzymes) both the sorption and desorption must be run with small flow rates, usually 10-50 ml/cm^2·h, to approximate a quasi-equilibrium regime in the column. It thus becomes necessary to have a strict quantitative estimate of the regime needed for quasi-equilibrium.

5.2. THEORETICAL PROBLEMS OF THE NON-EQUILIBRIUM DYNAMICS OF SORPTION AND CHROMATOGRAPHY INVOLVING SLOWLY DIFFUSING ORGANIC IONS IN IONITE GRAINS

The influence of kinetics on column ion-exchange may be evaluated from the non-equilibrium dynamics of sorption. The theory has been quite well developed now for one substance[372,374-386]. The mathematics of sorption dynamics (mass balance equations) can be formulated thus:

$$\varepsilon \frac{\partial C}{\partial t} + \frac{\partial <C>}{\partial t} = D_\ell \frac{\partial^2 C}{\partial x^2} - \frac{\partial C}{\partial x},$$ (5.25)

$$C(x,0) = 0, \quad 0 < x < \infty,$$ (5.26)

$$C(0,t) = C^0 = \text{const}, \quad 0 \leqslant t < \infty,$$ (5.27)

$$C(x,t) \text{ finite as } x \to \infty,$$ (5.28)

where C is the concentration of the substance in the solution,
    <C> is the average concentration in the sorbent per unit column volume,
    x is the distance from the outermost layer of sorbent in the column,
    t is time, and
    D_ℓ is the effective lengthwise migration coefficient.

The initial and boundary conditions are written in the form of (5.26)-(5.28) for frontal sorption from a solution of constant concentration. When describing elution chromatography, the following can be used instead of (5.27):

$$C(0,t) = \frac{M}{vS} \delta(t),$$ (5.29)

where $\delta(t)$ is the Dirac delta,
    S is the cross-sectional area of the column,
    v is the solution flow rate over the sorbent layer, and
    M is the quantity of substance in the column.

If a solution of the equation describing the kinetics of internal diffusion during sorption from a solution of constant concentration is known in the form of a function of the degree of completion F(t), then the kinetic relationship is conveniently given in the form

$$<C> = (1 - \varepsilon) \int_0^t \bar{C}^* (x,t - \xi) \frac{\partial F}{\partial \xi} d\xi = (1 - \varepsilon) \int_0^t F(t - \xi) \frac{\partial \bar{C}^*}{\partial \xi} d\xi,$$ (5.30)

where $\bar{C}^*(x,t)$ is the concentration on the surface of the bead which is defined by the sorption isotherm $\bar{C}^*(x,t) = K_d C$. Exact analytical solution of the dynamic problems is too cumbersome and inconvenient for analyzing real ion-exchange processes. It is much more convenient to apply the asymptotic solution for large and small periods. Instead of using the infinite series of F, the asymptotic forms for the short and long time periods can be substituted into (5.30), i.e., the forms for the regular and irregular regimes. For gel limitation kinetics we introduce the dimensionless parameters

$$\tau = (t - \frac{\varepsilon x}{v}) \frac{D}{\ell^2} \text{ (generalized time)} \tag{5.31}$$

and

$$\lambda = 3(1 - \varepsilon)(1 - \rho^3)K_d DX/\ell^2 v \text{ (generalized length)}, \tag{5.32}$$

where D is the diffusivity inside the bead, and for spherical beads $\ell = R$ and $\rho = 0$.

For frontal processes on spherical beads, we can obtain the asymptotic expressions

$$\frac{C}{C_0} = \ell^\lambda (1 - \frac{2}{\sqrt{\pi}} \int_0^{\lambda/2\sqrt{\tau}} \ell^{-\xi^2} d\xi) \tag{5.33}$$

for short times $t \leq R^2/15D$ and

$$\frac{C}{C_0} = \ell^{-2\lambda}[1 + \int_0^{2\pi\sqrt{2\lambda(\tau-0.13\lambda)}} \ell^{-\xi^2/8\lambda} I_1(\xi)d\xi] \tag{5.34}$$

for long periods $t \geq R^2/15D$, where $I_1(\xi)$ is a first-order Bessel function with an imaginary argument.

Solving the problem for gel limited kinetics and spherical beads shows that the elution curves in C-τ coordinates have the universal form for a given λ for the sorbent beads. Figure 5.4 shows the elution curves for frontal sorption calculated for various λ. It is easy to show that for $\lambda < 1$, when practically the whole of the curve can be described by Equation (5.34) – the irregular regime – the curves are very asymmetric, the x-co-ordinate of the turning point being well to the left (shorter times) of

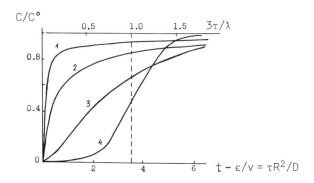

Fig. 5.4. Elution curves for dynamic sorption in different kinetic/dynamic regimes. Curves 1) $\lambda = 0.004$, $D = 10^{-11}$ cm^2/s; 2) $\lambda = 0.04$, $D = 10^{-10}$ cm^2/s; 3) $\lambda = 0.4$, $D = 10^{-9}$ cm^2/s; 4) $\lambda = 4.0$, $D = 10^{-8}$ cm^2/s, R = 0.01 cm. The dashed line shows the position of the equilibrium sharp front.

135

that of the line $\tau^{eq} = \lambda/3$, which corresponds to an equilibrium sharp front. As λ grows the undesirable effects weaken. For $\lambda > 4.5$, when the whole curve can be described by Equation (5.33) (the regular regime), the curve takes on the form characteristic of the quasi-equilibrium regime. Thus λ can serve as a criterion of the regularity of the regime when considering the sorption dynamics of spherical beads with internal diffusion kinetics.

If the isotherm is rectangular, it is also possible to obtain an analytical solution in the form of asymptotic approximations for long and short periods[16]. For the regular regime we get

$$\frac{C}{C_0} \sim 1 - \sum_{n=1}^{\infty} B_n e^{-\mu_n^2(\tau - \theta)}, \tag{5.35}$$

where θ is found from

$$\lambda = \phi(\theta) = 3\theta + \frac{1 + 3\rho + 6\rho^2 + 5\rho^3}{5(1 + \rho + \rho^2)}. \tag{5.36}$$

If the regime is irregular we get

$$\frac{C}{C_0} = -1 - \frac{2}{\pi} \arcsin \left[\sqrt{\pi/\tau} \cdot \frac{\lambda}{2(1 + \rho + \rho^2)}\right]. \tag{5.37}$$

It was shown in[16] that a dynamic regime is regularized 4.7 times faster for a rectangular isotherm than for a linear one, i.e., the quasi-equilibrium regime occurs for $\lambda > 1$. However, λ cannot serve as a regularity criterion if the beads are morphologically heterogeneous[16], i.e., for surface-layer and biporous ionites. A more general regularity criterion can be found using statistical moments. We have already shown that $\tau^{eq} = \lambda/3$ corresponds to the dimensionless time for a sharp equilibrium front to leave the column:

$$\tau^{eq} = (t^{eq} - t^0)\frac{D}{R^2} = (1 - \varepsilon)K_d \frac{D\kappa}{vR^2}. \tag{5.38}$$

Remembering the equation for the average sorption time in spherical beads (the first statistical moment of the kinetic curve $\bar{t}_1 = R^2/15D$), we obtain

$$\Lambda = \frac{t^{eq} - t^0}{5\bar{t}_1} \sim 1. \tag{5.39}$$

Since neither the bead radius nor the diffusivity appear explicitly in Equation (5.39), it is valid for surface-layer and biporous ionites, given internal diffusion kinetics.

An important parameter for frontal sorption and desorption is the degree of completion of the process. The theoretical frontal curves calculated from the theory for the kinetics of sorption [Equations (5.33) and (5.35)] can be used to calculate both the saturation and the yield in desorption for a column process η. In view of the asymptotic nature of the elution curves, it is necessary to set limits for the calculation, in either C-V or C-t coordinates, of the area corresponding to the sorbed or desorbed substance. A convenient boundary is the ratio C/C_0, i.e., the ratio of the approximate concentration of substance leaving the column to that entering. The boundary $1 - C/C_0$ can be taken for desorption, with ion exchange, since the sorption and desorption elution curves are symmetric. Figure 5.5 shows both the theoretical curves for $\eta - \lambda$ given

136

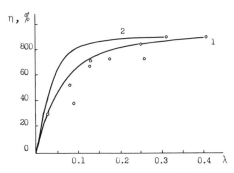

Fig. 5.5. Sorbent saturation η versus generalized length. The continuous
lines were calculated from theoretical curves for C/C_0 of 1)
0.90 and 2) 0.95. The points are the experimental data for
oxytetracycline on SBS-3 cationite for a C/C_0 leaving the column
of 0.9.

$C/C_0 = 0.95$ and $C/C_0 = 0.9$, and the experimental data for the degree of
completion of column sorptions of oxytetracycline on a sulfocationite.

5.3. INCREASING THE EFFICIENCY OF LOW-PRESSURE CHROMATOGRAPHY BY USING
      SURFACE-LAYER AND BIDISPERSED IONITES

     Our theoretical analysis of non-equilibrium sorption dynamics has
shown that it is possible on the whole to use three parameters to bring a
sorption or desorption regime to quasi-equilibrium, i.e., a column fully
saturated with sorbate, or a complete yield during desorption with a good
eluate concentration. These parameters are the solution flow rate, the
diffusivity of the ions in the ionite for gel limited rate of heterogeneous
mass transfer, and the length of the diffusion path of the ions in the
ionite. Making the columns longer must be done with caution. In Chapter
4 we considered the influence of height on the concentration of the
desorbate in the solution due to the formation of wide zones and how it is
possible to bring the eluate concentration close to the desorbate concentration in a dynamic frontal ion-exchange. However, our kinetic analysis
showed that the elution curves for sorption for given $\lambda(\Lambda)$ have the universal form in C-τ coordinates, but not in C-t coordinates. In real
processes, in which the latter and the analogous curves in C-V coordinates
are considered, the height of the column merely extends the ordinate of the
real elution curves. This can most easily be confirmed by considering the
height factor for the case of a linear isotherm, and the process has been
thoroughly studied both theoretically and experimentally. As the height of
the column is increased, both the frontal boundary and the eluate zones
become more spread, even though the dimensionless numbers λ and Λ both
grow. Thus in order to analyze the yield from a quasi-equilibrium regime,
the λ and Λ must be studied for particular column heights (e.g., 15-20 cm
for laboratory scale, 50-60 cm for pilot plant, and 1.5-2 m for industry).
The reduction of solution flow rate is not an efficient approach,
especially since prolonging the diffusion spreads the zone boundaries.

     We have also discussed the creation of ionites with the best perme-
abilities for organic ions. There are certain limitations. For practical
purposes the use of highly permeable, lightly crosslinked, and very swell-
able ionites in columns, especially tall ones, is not ideal, and in most
cases impossible. The crucial factor here is geometry, i.e., the length
of the diffusion path of the ions in the ionite. Since we have rejected
the use of small beads at high pressures, there are two rational approaches
to the problem, i.e., the use of surface-layer or bidispersed sorbents.

Bidispersed ionites differ from biporous sorbents in that the matrix in which the ionite particles are immobilized has a channel structure which does not hinder the diffusion of large ions.

Surface-layer ionites (Figure 5.6) are, in contrast to pellicular ionites[18], formed around an impermeable core. The thickness of the sorbing layer is defined by the Λ criterion for each system and consequently depends on V, D, h, and K. Calculations have shown that to sorb small organic ions such as amino acids or antibiotics with a diffusivity through the sorbing layer of about 10^{-8} cm^2/s in a quasi-equilibrium regime with fast flow rates, the layer should be about 20 μm. It is thus possible to get surface-layer ionites with beads 100-200 μm in diameter which are dense and hence hydrostatically stable when filtering through the sorbent layer. The ionites meet all our requirements and have quite a high exchange capacity (several tens of percent of the standard ionites).

The kinetics and dynamics of sorption in columns containing surface-layer ionites are carefully studied in[388-395]. Data on the dynamic desorption of lysine from a laboratory column containing the surface-layer sulfocationite SLC-10 are given in Table 5.5 together with data for a column filled with the gel sulfocationite KU-2 × 8 for comparison. The dynamic desorption in all the columns, except the first, is characterized by small λ and Λ. The first, which contains the SLC-10 ionite, has a Λ of 3.57, and hence a concentrated eluate with a complete yield can only be obtained from this ionite (Figure 5.7), this being especially clear in η-V coordinates, where η is the degree of completion. An analysis of the data in Table 5.5 shows, as would be expected, that the denser the cross-linking in the styrene-DVB copolymer, the less permeable the layer is to lysine when fixed to the glass sphere. It should be emphasized that the exchange capacity of SLC-10 is half that of a standard cationite.

Another feature of surface-layer materials can be seen from Table 5.6. This demonstrates how the diffusivity of oxytetracycline in a surface-layer cationite depends on the thickness of the layer. The effective diffusivity D_{eff} calculated assuming a final equilibrium loading of the layer by any organic counter-ion depends substantially on the layer thickness, although calculations done for various depths of counter-ion penetration have shown that if the layer is less than 25 μm the diffusivity remains constant. Thus we can assume that when thicker layers are formed the inner areas next

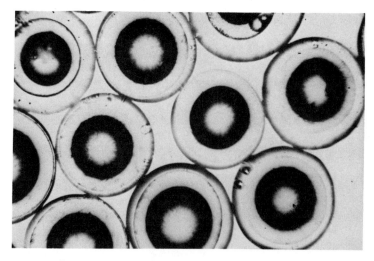

Fig. 5.6. Surface-layer cationites.

Table 5.5. Kinetic and Dynamic Parameters of the Desorption of Lysine from a Column Filled with Surface-Layer or KU-2 × 8 Gel Sulfocationite

Ionite	% DVB	Exchange capacity, mg-equ/g	R, μm	ℓ, μm	$D \cdot 10^8$, cm^2/s	λ	Λ
SLC-10*	10	2.40	90	30	2	–	3.57
SLC-15	15	2.05	88	28	0.82	–	0.82
SLC-20	20	1.67	86	26	0.19	–	0.62
KU-2 × 8	8	4.98	97	–	1.90	0.086	–

*SLC means surface-layer sulfocationite.
The solution flow rate was 400 ml/h·cm^2.

139

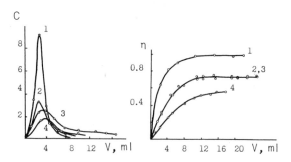

Fig. 5.7.   Desorption of lysine from surface-layer cationites. Curves
4) KU-2 × 8, and sulfocationites with 1) 10%, 2) 15%, and 3) 20%
DVB. Eluant rate 500 ml/h·cm².

Table 5.6.   Diffusion of Oxytetracycline on Surface-Layer Cationites with
Differing Sorbing Layer Thicknesses (18% crosslinking agent)

Ionite	R, μm	ℓ, μm	$D_{eff}·10^8$, cm²/s	$D·10^8$, cm²/s
SLC-1	62	19	0.24	0.24
SLC-2	96	53	0.55	0.23
SLC-3	122	79	1.08	0.29
SLC-4	175	132	1.41	0.20

to the core are compacted and inaccessible for organic ions. However, we
have shown that thicknesses of 20-25 μm are best for the quasi-equilibrium
dynamic processes involving organic ions, and the ionites' sorption
capacities are high enough for both preparative and quasi-equilibrium
processes to be economic.

Processes using surface-layer ionites can be significantly enhanced
and the solution flow rates increased tenfold over those usual for the
preparation of physiologically active substances on granulated ionite,
which we discussed in Chapter 4. For example, the desorption of oxytetracycline from a surface-layer ionite (Figure 5.8) can be implemented
using a flow rate of 600 ml/cm²·h to get a concentrated eluate with a full
yield. At these flow rates the zones would be spread on standard ionites.
Surface-layer carboxylic cationites can be used to get very good sorptive
preparations of complicated organic substances. Figure 5.9 shows the
elution curves for streptomycin and neomycin on the standard cationite
KB-4-P2. The apparent completion of the process indicated by the yield
concentration of the antibiotic approximating the feed one does not mean
the column is saturated, as this would be shown to be a dependence of the
degree on the volume of flowing solution. The true degree of completion
appears for long experiments with small flow rates. The opposite occurs
when the surface-layer ionite is used and the sorption is complete for a
limited quantity of solution even at very fast flow rates. The λ and Λ
criteria can be used to calculate the limits to the flow rates for which
the regime is close to quasi-equilibrium (Table 5.7). The theoretical and
experimental results in the table indicate that when the surface-layer
ionites are used the flow rates can be increased 20-fold.

Another advantage of these ionites can be seen for desorption (Figure
5.10). Complete yields of streptomycin and neomycin can only be achieved,
at fast flow rates, when these ionites are used. The calculations of the

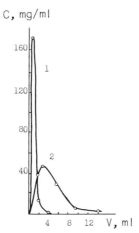

C, mg/ml

Fig. 5.8. Dynamic desorption of oxytetracycline from 1) surface-layer sulfocationite (4% DVB) and 2) Dowex-50 × 4 ionite. Eluant rate 600 ml/h·cm^2.

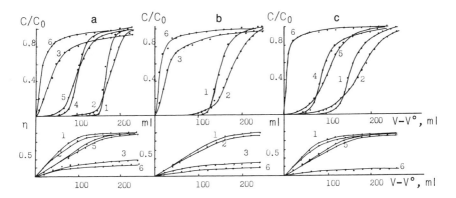

Fig. 5.9. Dynamic sorption of streptomycin (1, 2, 3) and neomycin (4, 5, 6) on surface-layer and standard ionites. Curves 1) and 4) surface-layer SLCI-17; 2) and 5) surface-layer SLCI-50; 3) and 6) standard KB-4P-2. Rate of mobile phase exiting column, ml/h·cm^2: a) 250; b) 500; c) 750.

Table 5.7. Limiting Flow Rates for the Dynamic Desorption of Neomycin, Streptomycin, and Kanamycin in Quasi-Equilibrium Regimes

Ionite	R, μm	ℓ, μm	Limiting solution flow rate, ml/h·cm^2		
			streptomycin	neomycin	kanamycin
SLCI-50	160	50	112	81	169
SLCI-17	49	17	487	325	552
KB-4P-2	325		4.2	3	5.5

141

flow rate limits for desorption (Table 5.8) show that the flow rate can be increased 100-fold on surface-layer ionites. We must emphasize that using rigid glass cores makes it possible to have very small overall diameters that do not hinder the flow, i.e., require an increase in pressure.

Some new avenues have been opened by bidispersed ionites ("Cellosorbent" is the trademark for a line of bidispersed ionites), which are particles of ionite in a permeable matrix (Figure 5.11) such that the mass of ionite may be comparable to that of the matrix. Organic ion diffusion is thus not hindered by the ionite grains. Studies of this sorbent type have found the effective diffusivities to be 10-100 times higher than those of the usual granulated ionites for the same sorbing material (Table 5.8).

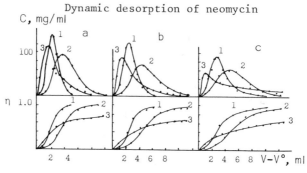

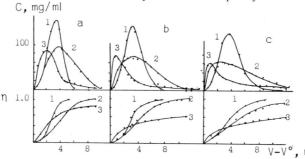

Fig. 5.10. Dynamic desorption of neomycin from 1) SLCI-17 and 2) SLCI-50 surface-layer ionites; and 3) KB-4P-2 standard ionite. Eluant 2 N ammonia; eluant rates, ml/h·cm²: a) 50; b) 100; and c) 250. Dynamic desorption of streptomycin from 1) SLCI-17 and 2) SLCI-50 surface-layer ionites; and 3) KB-4P-2 standard ionite. Eluant 1 N sulfuric acid; eluant rates, ml/h·cm²: a) 25; b) 50; and c) 100.

Table 5.8. Effective Diffusivities in Cellosorbent (a Bidispersed Ionite) and Its Granulated Standard Analogs

Counter-ion	Ionite	D_{eff}, cm²/s	
		standard ionite	Cellosorbent
Erythromycin	SNK-30D	$3.4 \cdot 10^{-9}$	$3.0 \cdot 10^{-8}$
Erythromycin	SNK-40H	$1.6 \cdot 10^{-8}$	$1.2 \cdot 10^{-7}$
Vitamin B_{12}	SG-1	$0.5 \cdot 10^{-9}$	$0.8 \cdot 10^{-7}$
Vitamin B_{12}	Biocarb-T	$3.5 \cdot 10^{-8}$	$2.4 \cdot 10^{-7}$
Oxytetracycline	KRS-5	$2.8 \cdot 10^{-9}$	$6.2 \cdot 10^{-8}$

Dynamic column processes bear out these kinetic parameters for oxytetracycline (Figure 5.12) and erythromycin (Figure 5.13). The λ criterion has been calculated as 2.2 for the desorption of the oxytetracycline for the regime shown, and 36.4 for the erythromycin. The corresponding figures for the standard granulated ionites are much worse, being less than 1. Thus the kinetic-dynamic criteria have been as useful for these ionites as they were for the surface-layer and standard ones. They enable us to judge different regimes in useful and convenient ways for the practical use of preparation processes in liquids, ion-exchange chromatography being one.

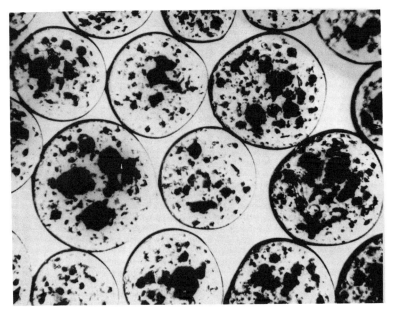

Fig. 5.11. Bidispersed sorbents with immobilized particles of ion-exchange resin.

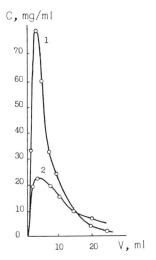

Fig. 5.12. Desorption of oxytetracycline from 1) sulfocellosorbent and 2) KRS-5P ionite. Eluant rate 250 ml/h·cm².

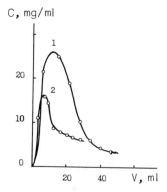

Fig. 5.13. Desorption of erythromycin from 1) sulfocellosorbent and 2) SNK-40H ionite. Eluant rate 20 ml/h·cm².

    The processes we have described above for separating and obtaining physiologically active substances under conditions in which sharp boundaries are formed between the ion zones can be scaled up so that they can be applied in industrial installations.

References

1. N. Hadden, F. Baumann, F. MacDonald, M. Munk, R. Stevenson, D. Gere, and F. Zamaroni, "Basic Liquid Chromatography," Varian Aerograph, U.S.A. (1971).
2. J. J. Kirkland (ed.), "Modern Practice of Liquid Chromatography," Wiley-Interscience, London (1971).
3. S. G. Perry, R. Amos, and P. I. Brewer, "Practical Liquid Chromatography," Plenum Press, London (1972).
4. H. Engelhardt, "Hochdruck-flussigkeitschromatographie," Springer-Verlag, Berlin (1977).
5. L. R. Snyder, and J. J. Kirkland, "Introduction to Modern Liquid Chromatography," John Wiley and Sons, New York (1979).
6. G. V. Samsonov, "Chromatography Applications in Biochemistry" (in Russian), Medgiz, Leningrad (1955).
7. G. V. Samsonov (ed.), "Antibiotic Separation and Purification. A Collection of Papers from the Leningrad Institute of Chemical Pharmacology" (in Russian), Goskhimizdat, Leningrad (1959).
8. G. V. Samsonov, "Antibiotic Sorption and Chromatography" (in Russian), AN SSSR, Moscow, Leningrad (1960).
9. G. V. Samsonov (ed.), "Physico-Chemical Methods of Studying, Analyzing, and Fractionating Biopolymers" (in Russian), Nauka, Moscow (1966).
10. "Selective Ion-Exchange Sorption of Antibiotics" (in Russian), Trudy Lenkhimfarminstituta, Leningrad (1968).
11. G. V. Samsonov, E. B. Trostyanskaya, and G. É. El'kin, "Ion Exchange. The Sorption of Organic Substances" (in Russian), Nauka, Leningrad (1969).
12. G. S. Libinson, "The Physico-Chemical Properties of Carboxylic Cationites" (in Russian), Nauka, Moscow (1969).
13. G. V. Samsonov and N. I. Nikitin (eds.), "Ion Exchange and Ionites" (in Russian), Nauka, Leningrad (1970).
14. G. V. Samsonov and P. G. Romankov (eds.), "Ionites and Ion Exchange" (in Russian), Nauka, Leningrad (1975).
15. G. S. Libinson, "The Sorption of Organic Compounds by Ionites" (in Russian), Meditsina, Moscow (1979).
16. L. K. Shataeva, N. N. Kuznetsova, and G. É. El'kin, "Carboxylic Cationites in Biology," G. V. Samsonov (ed.) (in Russian), Nauka, Leningrad (1979).
17. O. V. Orlievskaya, L. K. Shataeva, and G. V. Samsonov, Prikl.Biokhim.i Mikrobiol., 7:355 (1971).
18. C. Horvath, Pellicular ion-exchange resins in chromatography, in: "Ion Exchange and Solvent Extraction," J. Marinsky and Y. Marcus (eds.), Marcel Decker, Vol.5, p.207 (1973).
19. A. M. Elyashevich, Polymer, 20:1383 (1979).
20. R. Kunin and K. A. Kun, J.Polym.Sci., Part C, No.16, p.1457 (1967).
21. D. G. Griffin, Ind.Eng.Chem., 44:2686 (1952).

22. V. L. Vel'dyaskina and V. N. Trushin, "Chemistry and Chemical Technology" (in Russian), Trudy Kuzbasskogo Politekh.Inst., No.36, p.134, Kemerovo (1972).

23. A. B. Pashkov, M. I. Itkina, E. I. Lyustgarten, V. A. Grigor'ev, R. R. Dranovskaya, S. M. Tkachuk, T. K. Brutskus, A. M. Prokhorova, and S. K. Vinogradov, "Production and Treatment of Plastics and Synthetic Resins" (in Russian), Trudy NIIPM, Moscow, No.7, p.2 (1973).

24. V. N. Trushin, "Production and Treatment of Plastics and Synthetic Resins" (in Russian), Trudy NIIPM, Moscow, p.3 (1973).

25. V. A. Davankov, S. V. Rogozhin, and M. P. Tsyurupa, Vysokomol.Soed., Seriya B, 15:463 (1973).

26. V. A. Davankov, M. P. Tsyurupa, and S. V. Rogozhin, Angewandte Markromol.Chem., 32:145 (1973).

27. V. A. Davankov, S. V. Rogozhin, M. P. Tsyurupa, and E. A. Pankratov, Zh.Fiz.Khim., 48:2964 (1974).

28. V. A. Davankov, S. V. Rogozhin, and M. P. Tsyurupa, J.Polym.Sci., Polym.Symp., 47:95 (1974).

29. M. P. Tsyurupa, V. A. Davankov, and S. V. Rogozhin, ibid., p.189.

30. V. A. Davankov, S. V. Rogozhin, and M. P. Tsyurupa, in: "Ion Exchange and Solvent Extraction," Vol.7, New York, p.29 (1977).

31. J. R. Millar, D. C. Smith, W. E. Marr, and T. R. E. Kressman, J.Chem.Soc., p.218 (1963).

32. J. R. Millar, D. C. Smith, and T. R. E. Kressman, J.Chem.Soc., p.304 (1965).

33. K. A. Kunin and R. Kun, J.Polym.Sci., A-I, 6:2689 (1968).

34. J. Scide, J. Malinsky, K. Dusek, and W. Heits, Adv.in Polym.Sci., 5:113 (1967).

35. A. A. Tager, M. V. Tsilipotkina, E. B. Makovskaya, A. B. Pashkov, E. I. Lyustgarten, and M. A. Pechenkina, Vysokomol.Soed.,Seriya A., p.1065 (1968).

36. J. Coupek, M. Krivakova, and S. Pokorny, J.Polym.Sci.,Polym.Symp., 42:185 (1973).

37. K. M. Saldadze and V. D. Kopylova-Valova, "Complex-Forming Ionites (Complexites)" (in Russian), Khimiya, Moscow (1980).

38. W. L. Sederel and G. J. DeJong, J.Appl.Polym.Sci., 17:2835 (1973).

39. H. Jacobelli, M. Bartholin, and A. Guyot, J.Appl.Polym.Sci., 23:927 (1979).

40. A. A. Tager, M. V. Tsilipotkina, E. B. Makovskaya, E. I. Lyustgarten, A. B. Pashkov, and M. A. Lagunova, Vysokomol.Soed.,Seriya A., 13:2370 (1971).

41. T. R. E. Kressman and I. R. Millar, English Patent No. 860695 (1961).

42. A. A. Tager and M. V. Tsilipotkina, Uspekhi Khimii, 47:No.1, p.152 (1978).

43. J. Spevacek, and K. Dusek, J.Polym.Sci., 18:No.10, p.2027 (1980).

44. M. Sova and Z. Pelzbauer, Coll.Czech.Chem.Communications, 43:No.7, p.1677 (1978).

45. V. A. Dinaburg, G. V. Samsonov, K. M. Genender, V. A. Pasechnik, V. S. Yurchenko, G. E. El'kin, and S. F. Belaya, Zh.Prikl.Khim., 41:891 (1968).

46. G. V. Samsonov, V. A. Dinaburg, V. A. Pasechnik, G. E. El'kin, V. S. Yurchenko, S. F. Belaya, and K. M. Genender, Zh.Prikl.Khim., 44:859 (1971).

47. K. B. Musabekov, V. A. Dinaburg, and G. V. Samsonov, Zh.Prikl.Khim., 42:82 (1969).

48. E. E. Ergozhin and M. Kurmanaliev, Vestnik AN Kaz.SSR, No.9, p.53 (1971).

49. O. P. Kolomeitsev, N. N. Kuznetsova, and V. A. Dinaburg, in: "Ion Exchange and Ionites" (in Russian), Nauka, Leningrad, p.48 (1970).

50. O. P. Kolomeitsev, V. A. Pasechnik, N. N. Kuznetsova, and G. V. Samsonov, Vysokomol.Soed.,Seriya A, 14:1746 (1972).

51. V. S. Yurchenko, K. P. Papukova, N. N. Kuznetsova, B. V. Moskvichev, and G. V. Samsonov, Zh.Prikl.Khim., 46:2779 (1973).

52. S. F. Belaya, G. E. El'kin, and G. V. Samsonov, Kolloidn.Zh., 33:No.5, p.645 (1971).

53. A. Sh. Genedi and G. B. Samsonov, Zh.Fiz.Khim., 44:No.12, p.3128 (1970).

54. A. A. Vasil'ev, "Synthesis of Polymeric Insoluble Sulfonic Acids. Sulfoacid Ionites" (in Russian), Nauka, Leningrad (1971).

55. E. E. Ergozhin and M. Kurmanaliev, Trudy Inst.Khim.Nauk AN Kaz.SSR, Alma-Ata, 32:34 (1972).

56. E. E. Ergozhin, "Very Permeable Ionites" (in Russian), Nauka, Kaz. SSR, Alma-Ata (1979).

57. G. V. Samsonov and A. B. Petushina, "Antibiotic Separation and Purification" (in Russian), Trudy Lenkhimfarminstituta, Goskhimizdat, Leningrad, p.28 (1959).

58. A. A. Selezneva, N. I. Dubinina, O. F. Luknitskaya, and G. V. Samsonov, Kolloidn.Zh., 38:1194 (1976).

59. G. Ya. Gerasimova, L. F. Yakhontova, and B. P. Bruns, "Investigations in the Field of Industrial Sorbents" (in Russian), AN SSSR, Moscow, p.70 (1961).

60. G. S. Libinson, E. M. Savitskaya, and B. P. Bruns, Vysokomol.Soed., 2:No.10, p.1500 (1960).

61. P. S. Porter and M. J. Semmens, Environment International, 3:311 (1980).

62. T. I. Pristoupil, M. Kramlova, M. Kubin, and P. Spacec, J.Chromatogr., 67:362 (1972).

63. V. S. Soldatov, G. V. Gorbunov, S. B. Makarova, and S. I. Bezuevskaya, Zh.Prikl.Khim., 53:2559 (1980).

64. K. I. Radzyavichus, L. K. Shataeva, G. V. Samsonov, N. G. Zhukova, A. I. Zorina, and B. N. Laskorin, Vysokomol.Soed.,Seriya A, 24:1066 (1982).

65. S. Fisher and R. Kunin, J.Phys.Chem., 60:1031 (1956).

66. V. S. Yurchenko, V. A. Pasechnik, N. N. Kuznetsova, K. M. Rozhetskaya, L. Ya. Solov'eva, and G. V. Samsonov, Vysokomol.Soed.,Seriya A, 21:179 (1979).

67. J. Baldrian, B. N. Kolarz, and H. Galina, Collection Czechoslovak Chemical Communications, 46:1675 (1981).

68. K. Kun and R. Kunin, J.Polym.Sci., B-2, p.389 (1964).

69. H. Hilgen, G. J. DeJong, and W. L. Sederel, J.Appl.Polym.Sci., 19:2647 (1975).

70. D. M. Freifelder, "Physical Biochemistry," Freeman, San Francisco (1976).

71. A. A. Vansheidt, A. A. Vasil'ev, and O. I. Okhrimenko, "Chromatography" (in Russian), Lengosuniversiteta, Leningrad, p.52 (1956).

72. I. A. Chernova and G. V. Samsonov, Vysokomol.Soed.,Seriya A, 21:1608 (1979).

73. V. F. Colombo and P. Y. Späth, Int.J.Scann.Electron Microsc.Relat.Techn.and Appl., No.3, p.515 (1981).

74. A. A. Éfendiev and A. T. Shakhtakhtinskaya, Vysokomol.Soed.,Seriya A, 20:No.2, p.314 (1978).

75. V. S. Soldatov, "Simple Ion-Exchange Equilibria" (in Russian), Nauka i Tekhnika, Minsk (1968).

76. S. J. Gregg and K. S. W. Sing, "Adsorption, Surface Area and Porosity," Academic Press (1967).

77. Z. Peltsbauer and V. Forst, Collection Czechoslovak Chemical Communications, 31:2338 (1966).

78. M. M. Dubinin (ed.), "Methods for Investigating the Structures of Very Dispersed and Porous Bodies" (in Russian), AN SSSR, Moscow, pp.164, 178 (1958).

79. K. Kunin and R. Kun, J.Polym.Sci., C-16:1457 (1967).

80. E. B. Trostyanskaya, A. S. Tevlina, and F. A. Naumova, Vysokomol.Soed., 5:1240 (1963).
81. D. C. Havard and R. Wilson, J.Colloid. and Interface Sci., 57:276 (1976).
82. M. V. Tsilipotkina, "Modern Physical Methods for Investigating Polymers" (in Russian), Khimiya, Moscow, p.198 (1982).
83. M. M. Dubinin, "Modern Capillary Theory" (in Russian), Khimiya, Leningrad, p.100 (1980).
84. V. V. Boginskii, Z. Ya. Chernousova, G. A. Akopov, G. Ya. Isaeva, and N. G. Polyanskii, Zh.Fiz.Khim., 46:1871 (1972).
85. T. G. Plachenov, in: "Adsorption and Porosity" (in Russian), Nauka, Moscow, p.194 (1976).
86. J. Baldrian, J. Plestil, and J. Stamberg, Collection Czechoslovak Chemical Communications, 41:3555 (1976).
87. W. O. Statton, in: "Newer Methods of Polymer Characterization," B. Ke (ed.), Wiley-Interscience (1964).
88. N. N. Kuznetsova, V. S. Yurchenko, K. P. Papukova, T. D. Murav'eva, N. I. Dubinina, L. M. Tsukanova, and G. V. Samsonov, Vysokomol.Soed., Seriya B, 21:244 (1979).
89. Yu. S. Nadezhin, L. K. Shataeva, N. N. Kuzetsova, A. V. Sidorovich, and G. V. Samsonov, Vysokomol.Soed.,Seriya A, 17:448 (1975).
90. S. B. Makarova, Zh. M. Litvak, I. A. Vakhtina, and E. V. Egorov, Vysokomol.Soed.,Seriya A, 13:2160 (1971).
91. G. W. Longman, G. D. Wignal, M. Hemming, and J. V. Dawkings, Colloid and Polymer Sci., 252:298 (1974).
92. L. Z. Vilenchik, O. I. Kurenbin, T. P. Zhmakina, and T. P. Belen'kii, Dokl.AN SSSR, 250:381 (1980).
93. L. Z. Vilenchik, O. I. Kurenbin, T. P. Zhmakina, V. S. Yurchenko, and B. G. Belen'kii, Zh.Fiz.Khim., 55:182 (1981).
94. E. F. Casassa, J.Polym.Sci., 5:773 (1967).
95. I. A. Chernova, T. E. Pogodina, L. K. Shataeva, and G. V. Samsonov, Vysokomol.Soed.,Seriya A, 22:2403 (1980).
96. H. Hradil, Angewandte Makromol.Chem., 66:51 (1978).
97. G. E. El'kin, S. F. Klikh, and A. T. Melenevskii, Zh.Prikl.Khim., 50:812 (1977).
98. J. S. Redirha and J. A. Kitchener, Trans.Faraday Soc., 59:515 (1963).
99. N. I. Nikolaev, in: "The Kinetics and Dynamics of Physical Adsorption" (in Russian), Nauka, Moscow (1973).
100. N. I. Nikolaev, A. M. Filimonova, and N. N. Tunitskii, Zh.Fiz.Khim., 43:1249 (1969).
101. T. R. F. Kressman and J. A. Kitchener, Disc.Faraday Soc., 7:90 (1949).
102. D. K. Hale and D. Reichenberg, Disc.Faraday Soc., 7:79 (1949).
103. D. Reichenberg, J.Am.Chem.Soc., 75:589 (1953).
104. Yu. A. Kokotov and V. A. Pasechnik, "Ion-Exchange Equilibrium and Kinetics" (in Russian), Khimiya, Leningrad (1970).
105. F. Helfferich, Ionenaustauscher B. I. Grundlagen. Structur – Herstellung Theorie. Verlag Chemic – GMBH – Weinheim Bergstr. (1959).
106. A. W. Adamson and G. Grossman, J.Chem.Phys., 17:1002 (1949).
107. G. E. Boyd, A. Adamson, and G. E. Myers, J.Am.Chem.Soc., 69:2836 (1947).
108. N. N. Tunitskii, V. A. Kaminskii, and S. F. Timashev, "Methods of Physico-Chemical Kinetics" (in Russian), Khimiya, Moscow, p.197 (1972).
109. G. S. Libinson, E. M. Savitskaya, and B. P. Bruns, Zh.Fiz.Khim., 37:420,641 (1963).
110. E. S. Vaisberg, L. F. Yakhontova, and B. P. Bruns, Zh.Fiz.Khim., 41:892 (1967).
111. S. F. Belaya, G. E. El'kin, and G. V. Samsonov, "Selective Ion-Exchange Sorption of Antibiotics" (in Russian), Trudy Lenkhimfarminstituta, Leningrad, p.147 (1968).

112. A. A. Selezneva, N. I. Dubinina, G. A. Babenko, O. F. Luknitskaya, and G. V. Samsonov, Kolloidn.Zh., 37:No.6, p.1138 (1975).
113. G. A. Yaskovich, E. P. Kozneva, G. É. El'kin, V. Ya. Vorob'eva, and G. V. Samsonov, Khim.-farm.Zh., No.10, p.47 (1974).
114. G. É. El'kin, G. A. Babenko, A. A. Selezneva, and G. V. Samsonov, Kolloidn.Zh., 34:No.2, p.208 (1972).
115. L. K. Shataeva and G. V. Samsonov, Khim.-farm.Zh., 11:No.4, p.78 (1977).
116. A. A. Selezneva and G. V. Samsonov, Khim.-farm.Zh., No.7, p.77 (1981).
117. G. V. Samsonov, Pure and Appl.Chem., 38:151 (1974).
118. P. H. Geil, J.Macromolec.Sci.Phys., B-12:173 (1976).
119. A. A. Berlin, Vysokomol.Soed.,Seriya A, 20:483 (1978).
120. A. Katchalsky and S. Lifson, J.Polym.Sci., 11:409 (1953).
121. G. V. Samsonov and L. L. Bosak, Meditsinskaya Prom.SSSR, No.7, p.32 (1962).
122. G. V. Samsonov, V. V. Vedeneeva, and A. A. Selezneva, Dokl.AN SSSR, 125:591 (1959).
123. R. Kunin, "Ion Exchange in the Process Industry," London Soc. of Chem. Ind., p.10 (1970).
124. G. V. Samsonov and T. I. Lesment, "Antibiotic Separation and Purification" (in Russian), Trudy Lenkhimfarminstituta, Goskhimizdat, Leningrad, p.67 (1959).
125. Z. Pel'tsbauer, in: "Physico-Chemical Properties and Synthesis of Macromolecular Compounds" (in Russian), Naukova Dumka, Kiev, p.76 (1976).
126. L. M. Barclay, Angewandte Makromol.Chem., 52:1 (1976).
127. T. D. Murav'eva, Vysokomol.Soed.,Seriya B, 24:163 (1982).
128. K. Dusek, J.Polym.Sci., B-3:209 (1965).
129. K. Dusek, J.Polym.Sci., C-16:1289 (1967).
130. K. Dusek, in: "Polymer Networks Structural and Mechanical Properties," A. J. Chompff and S. Newman (eds.), Plenum, London, p.245 (1971).
131. H. Jacobelli, M. Bartholin, and A. Guyot, J.Appl.Polymer Sci., 23:927 (1979).
132. G. V. Samsonov, A. A. Khints, and V. P. Solomatina, Antibiotiki, No.6, p.27 (1958).
133. A. Katchalsky, Progr.Biophys.,Biophys.Chem., 4:1 (1954).
134. A. Katchalsky and J. Michaeli, J.Polym.Sci., 15:69 (1955).
135. J. Michaeli and A. Katchalsky, J.Polym.Sci., 23:683 (1957).
136. H. P. Gregor, J.Am.Chem.Soc., 70:1293 (1948); 73:642 (1951).
137. V. A. Irzhak, B. A. Rozenberg, and N. S. Enikolopyan, "Network Polymers" (in Russian), Nauka, Moscow (1979).
138. I. F. Khirsanova, A. I. Pokrovskaya, V. S. Soldatov, and V. A. Artamonov, Izv. AN BSSR, Seriya Khim., p.19 (1974).
139. T. K. Brutskus, K. M. Saldadze, É. A. Uvarova, and E. I. Lyustgarten, Kolloidn.Zh., 35:445 (1973).
140. K. Dusek, H. Galina, and J. Mikes, Polym.Bull., 3:19 (1980).
141. B. Jerslev, L. Brehm, and M. Vahl Gabrielsen, Acta Chem.Scand., B-31:875 (1977).
142. D. Horak, Z. Pelzbauer, M. Bleha, M. Ilavsky, F. Svec, and J. Kalal, J.Appl.Polym.Sci., 26:411 (1981).
143. N. N. Kuznetsova, K. M. Rozhetskaya, B. V. Moskvichev, L. K. Shataeva, A. A. Selezneva, I. M. Ogorodnova, and A. V. Selezneva, Vysokomol.Soed.,Seriya A, 18:355 (1976).
144. R. L. Gustavson and J. A. Lirio, J.Phys.Chem., 69:2849 (1965).
145. G. V. Samsonov and A. P. Bashkovich, "Antibiotic Separation and Purification" (in Russian), Trudy Lenkhimfarminstituta, Goskhimizdat, Leningrad, p.57 (1959).
146. N. N. Gavrilova, V. S. Pirogov, A. D. Morozova, N. N. Kuznetsova, and G. V. Samsonov, Zh.Fiz.Khim., 54:468 (1980).
147. A. Sh. Genedi and G. V. Samsonov, "Selective Ion-Exchange Sorption of

Antibiotics" (in Russian), Trudy Lenkhimfarminstituta, Leningrad, p.164 (1968).

148. V. S. Yurchenko, G. I. Kil'fin, K. B. Musabekov, V. A. Pasechnik, and G. V. Samsonov, "Selective Ion-Exchange Sorption of Antibiotics" (in Russian), Trudy Lenkhimfarminstituta, Leningrad, p.121 (1968).

149. L. V. Dmitrenko, D. I. Ostrovskii, and G. V. Samsonov, "Work from the Second All-Union Symposium on Ion-Exchange Thermodynamics" (in Russian), Minsk, p.42 (1975).

150. L. M. Ovsyanko, T. L. Starobinets, and M. Vinarskii, Izv. AN BSSR, Seriya Khim.Nauk, No.3, p.36 (1973).

151. B. V. Moskvichev, V. S. Yurchenko, A. Sh. Genedi, B. Sh. Chokina, and G. V. Samsonov, "Ion Exchange and Ionites" (in Russian), Nauka, Leningrad, p.142 (1970).

152. B. V. Moskvichev, V. S. Yurchenko, A. Sh. Genedi, B. Sh. Chokina, and G. V. Samsonov, "Synthesis, Structure, and Properties of Polymers" (in Russian), Nauka, Leningrad, p.263 (1970).

153. B. N. Trushin, A. B. Davankov, and V. V. Korshak, Vysokomol.Soed., Seriya A, 9:1140 (1967).

154. B. N. Trushin and A. B. Davankov, Sbornik Nauchnykh Trudov Kuzbasskogo Politekhnicheskogo Instituta, No.26, p.226 (1970).

155. A. B. Davankov and A. B. Zubakova, Vysokomol.Soed., 2:884 (1960).

156. A. B. Davankov and O. A. Vitols, Vysokomol.Soed., 4:1093 (1962).

157. L. K. Arkhangel'skii and E. A. Materova, "Physico-Chemical Properties of Solutions" (in Russian), Izdat. Lengosuniversiteta, Leningrad, p.163 (1964).

158. G. V. Samsonov and V. A. Pasechnik, Uspekhi Khimii, 38:1257 (1969).

159. L. F. Yakhontova, E. M. Savitskaya, and B. P. Bruns, "Chromatography, Theory and Applications" (in Russian), AN SSSR, Moscow, p.100 (1960).

160. H. P. Gregor, B. R. Sundheim, K. M. Held, and M. H. Waxman, J.Colloid Sci., 7:511 (1952).

161. B. R. Sundheim, M. H. Waxman, and H. P. Gregor, J.Phys.Chem., 57:974 (1953).

162. G. E. Boyd and B. A. Soldano, Z. Electrochem., 57:162 (1953).

163. S. Lapanje and D. Dolar, Z.Phys.Chem., 18:11 (1958); 21:376 (1959).

164. K. B. Musabekov, V. A. Pasechnik, and G. V. Samsonov, Zh.Fiz.Khim., 44:991 (1970).

165. V. A. Pasechnik, K. B. Musabekov, and G. V. Samsonov, Zh.Prikl.Khim., 56:76 (1973).

166. N. N. Gavrilova, V. S. Pirogov, A. D. Morozova, Yu. S. Nadezhin, N. N. Kuznetsova, and G. V. Samsonov, Zh.Prikl.Khim., 54:1190 (1981).

167. B. P. Nikol'skii and V. I. Paramonova, Uspekhi Khimii, 8:1535 (1939).

168. V. S. Soldatov, "Ion Exchange" (in Russian), Nauka, Moscow, p.111 (1981).

169. V. A. Pasechnik, K. B. Musabekov, and G. V. Samsonov, "Synthesis, Structure, and Properties of Polymers" (in Russian), Nauka, Leningrad, p.265 (1970).

170. E. V. Anufrieva and Yu. Ya. Gotlib, Adv. in Polym.Sci., 40:3 (1981).

171. M. G. Krakovyak, S. S. Anufrieva, and S. S. Skorokhodov, Vysokomol. Soed.,Seriya A, 11:2499 (1969).

172. E. V. Anufrieva, N. P. Kuznetsova, M. G. Krakovyak, R. N. Mishaeva, V. D. Pautov, G. V. Semisotnov, and T. V. Sheveleva, Vysokomol.Soed., Seriya A, 19:102 (1977).

173. E. V. Anufrieva, T. M. Birshtein, T. N. Nekrasova, O. B. Ptytsyn, and T. V. Sheveleva, J.Polym.Sci., C-16:3519 (1968).

174. N. P. Kuznetsova, R. N. Mishaeva, L. R. Gudkin, E. V. Anufrieva, V. D. Pautov, and G. V. Samsonov, Vysokomol.Soed.,Seriya A, 19:107 (1977).

175. A. R. Mathieson and R. T. Shet, J.Polym Sci., A-4:2945 (1966).

176. J. C. Leyte and M. Mandel, J.Polym.Sci., A-2:1879 (1964).

177. T. N. Nekrasova, E. V. Anufrieva, A. M. El'yashevich, and O. B.

Ptitsyn, Vysokomol.Soed.,Seriya A, 7:915 (1965).

178. Z. S. Tabidze, L. F. Yakhontova, B. P. Bruns, and K. M. Saldadze, Plastmassy, No.3, p.333 (1963).

179. A. D. Morozova, V. S. Pirogov, T. V. Chervyak, and G. V. Samsonov, Zh.Prikl.Khim., 51:327 (1978).

180. L. V. Dmitrenko, A. D. Morozova, and G. V. Samsonov, Izv.AN SSSR, Seriya Khim., No.7, p.1563 (1972).

181. N. N. Kuznetsova, K. P. Papukova, T. D. Murav'eva, G. V. Bilibina, and V. N. Andreeva, Vysokomol.Soed.,Seriya A, 20:1957 (1978).

182. L. V. Dmitrenko, V. S. Pirogov, and G. V. Samsonov, "Ion-Exchange Thermodynamics" (in Russian), Nauka i Tekhnika, Minsk, p.162 (1968).

183. G. V. Samsonov and N. P. Kuznetsova, Dokl.AN SSSR, 115:351 (1957).

184. N. P. Kuznetsova, R. N. Mishaeva, N. N. Kuznetsova, K. M. Rozhetskaya, and G. V. Samsonov, Vysokomol.Soed.,Seriya B, 22:874 (1980).

185. G. E. Dusek and G. R. Stark, Proc.Nat.Acad.Sci.,USA, 66:651 (1970).

186. N. P. Kuznetsova, R. N. Mishaeva, L. R. Gudkin, N. N. Kuznetsova, T. D. Murav'eva, K. P. Papukova, K. M. Rozhetskaya, and G. V. Samsonov, Vysokomol.Soed.,Seriya A, 20:629 (1978).

187. A. Katchalsky and P. Spitnik, J.Polym.Sci., 2:432 (1947).

188. E. V. Anufrieva, V. D. Pautov, N. P. Kuznetsova, and R. N. Mishaeva, Vysokomol.Soed.,Seriya B, 23:557 (1981).

189. P. S. Nys and E. M. Savitskaya, Zh.Fiz.Khim., 43:1536 (1969).

190. E. M. Savitskaya, N. L. Shelenberg, and P. S. Nys, Antibiotiki, No.4, p.294 (1966).

191. D. Reichenberg and D. J. McCanley, J.Chem.Soc., p.2741 (1955).

192. E. Glueckauf, Proc.Roy.Soc., A-258:350 (1962).

193. O. D. Bonner and V. F. Holland, J.Phys.Chem., 60:1102 (1956).

194. E. Högfeldt, F. Ekedahl, and J. G. Sillen, Acta Chem.Scand., 4:828 (1950).

195. V. S. Soldatov, A. I. Pokrovskaya, and R. V. Martsinkevich, Zh.Fiz.Khim., 41:1098 (1967).

196. B. R. Spinner, J. Ciric, and W. J. Graydon, Canad.J.Chem., 32:143 (1954).

197. E. Högfeldt, Arkiv for Kemi (Sweden), 13:491 (1958).

198. L. K. Arkhangel'skii, "Ion-Exchange Thermodynamics" (in Russian), Nauka i Tekhnika, Minsk, p.49 (1968).

199. L. F. Yakhontova, E. M. Savitskaya, and B. P. Bruns, "Investigations of Ion-Exchange, Distributive, and Precipitation Chromatography" (in Russian), Izd.AN SSSR, p.3 (1959).

200. J. Feitelson, Arch.Biochem.Biophys., 79:177 (1969).

201. W. Argersinger, A. W. Davidson, and O. D. Bonner, Trans.Kansas Acad.Sci., 53:404 (1950).

202. N. A. Izmailov and S. Kh. Mushinskaya, Zh.Fiz.Khim., 36:1210 (1962).

203. G. L. Gains and H. C. Thomas, J.Chem.Phys., 21:No.4, p.714 (1953).

204. T. R. E. Kressman and J. A. Kitchener, J.Chem.Soc., p.1208 (1949).

205. R. Lamry and F. Raienber, Biopolymers, 9:1125 (1970).

206. E. H. Cruikshank and P. Mears, Trans.Faraday Soc., 53:1289 (1957).

207. G. V. Samsonov and N. N. Momot, Zh.Prikl.Khim., 45:1383 (1972).

208. L. V. Naumova, T. V. Anuchina, V. Ya. Vorob'eva, and G. V. Samsonov, Khim.-farm.Zh., 11:No.10, p.108 (1977).

209. G. V. Samsonov and L. V. Dmitrenko, "Ion-Exchange Thermodynamics" (in Russian), Nauka i Tekhnika, Minsk, p.178 (1968).

210. E. M. Savitskaya, L. F. Yakhontova, and P. S. Nys, "Ion Exchange" (in Russian), Nauka, Moscow, p.229 (1981).

211. L. V. Dmitrenko, K. K. Kalnin'sh, V. Ya. Vorob'eva, V. G. Belen'kii, and G. V. Samsonov, "Selective Ion-Exchange Sorption of Antibiotics" (in Russian), Trudy Lenkhimfarminstituta, Leningrad, p.20 (1968).

212. K. J. Brunings and R. V. Wandward, J.Am.Chem.Soc., 78:4155 (1956).

213. G. V. Samsonov, A. P. Bashkovich, V. G. Gvozdeva and L. A. Moiseenko, "Problems of Fermentation and Purification of Antibiotics" (in

Russian), Trudy Lenkhimfarminstituta, Leningrad, p.185 (1962).

214. G. V. Samsonov and V. Ya. Vorob'eva, "Problems of Fermentation and Purification of Antibiotics" (in Russian), Trudy Lenkhimfarm-instituta, Leningrad, p.191 (1962).

215. V. Ya. Vorob'eva and G. V. Samsonov, Antibiotiki, No.1, p.73 (1963).

216. A. P. Bashkovich and G. V. Samsonov, Kolloidn.Zh., 26:613 (1964).

217. E. I. Gurina, G. E. El'kin, L. V. Dmitrenko, and G. V. Samsonov, "Selective Ion-Exchange Sorption of Antibiotics" (in Russian), Trudy Lenkhimfarminstituta, Leningrad, p.29 (1968).

218. L. V. Dmitrenko, A. Sh. Genedi, and G. V. Samsonov, Kolloidn.Zh., 32:37 (1970).

219. B. V. Moskvichev, A. Sh. Genedi, and G. V. Samsonov, Zh.Prikl.Khim., 43:1171 (1970).

220. L. V. Dmitrenko, N. D. Lipinskaya, and G. V. Samsonov, Kolloidn.Zh., 33:670 (1971).

221. L. V. Dmitrenko, A. D. Morozova, V. S. Pirogov, and G. V. Samsonov, Vysokomol.Soed.,Seriya B, 24:353 (1972).

222. L. V. Dmitrenko, E. B. Vulikh, and G. V. Samsonov, Izv.AN SSSR,Seriya Khim., No.3, p.730 (1972).

223. A. D. Morozova, L. V. Dmitrenko, V. S. Pirogov, and G. V. Samsonov, Izv.AN SSSR,Seriya Khim., No.8, p.1747 (1972).

224. G. V. Samsonov, O. P. Kolomeitsev, T. V. Atabekyan, V. Ya. Vorob'eva, and V. V. Azanova, "Ionites and Ion Exchange" (in Russian), Nauka, Leningrad, p.134 (1975).

225. V. Ya. Vorob'eva, L. A. Selezneva, and N. V. Posessor, Khim.-farm.Zh., No.12, p.31 (1972).

226. G. V. Samsonov, B. V. Moskvichev, and N. N. Kuznetsova, "Ion Exchange and Ionites" (in Russian), Nauka, Leningrad, p.90 (1970).

227. B. V. Moskvichev, N. N. Kuznetsova, and G. V. Samsonov, Zh.Fiz.Khim., 45:1873 (1971).

228. E. H. Flynn and M. V. Sigal, J.Am.Chem.Soc., 76:3121 (1954).

229. M. Gerol'd, "Antibiotics" (in Russian), Meditsina, Moscow (1966).

230. L. K. Shataeva and G. V. Samsonov, "Ion-Exchange Thermodynamics" (in Russian), Nauka i Tekhnika, Minsk, p.193 (1968).

231. V. Ya. Vorob'eva, M. V. Kravets, and G. V. Samsonov, Kolloidn.Zh., 32:427 (1970).

232. L. K. Shataeva and G. V. Samsonov, "Selective Ion-Exchange Sorption of Antibiotics" (in Russian), Trudy Lenkhimfarminstituta, Leningrad, p.42 (1968).

233. S. F. Klikh and G. V. Samsonov, "Selective Ion-Exchange Sorption of Antibiotics" (in Russian), Trudy Lenkhimfarminstituta, Leningrad, p.55 (1968).

234. L. K. Shataeva and G. V. Samsonov, Antibiotiki, No.10, p.867 (1969).

235. N. S. Rakutina, V. A. Borisova, K. P. Papukova, N. N. Nemtsova, G. E. El'kin, V. S. Pirogov, and G. V. Samsonov, Zh.Prikl.Khim., 55:540 (1982).

236. N. S. Rakutina, M. Kh. Rubene, G. E. El'kin, and G. V. Samsonov, Zh.Prikl.Khim., 55:780 (1982).

237. N. S. Rakutina, T. A. Shapenyuk, G. E. El'kin, and G. V. Samsonov, Zh.Prikl.Khim, 55:1178 (1982).

238. G. V. Samsonov, V. V. Vedeneeva, V. V. Shatik, and T. A. Vikhoreva, "Problems of Fermentation and Purification of Antibiotics" (in Russian), Trudy Lenkhimfarminstituta, Leningrad, p.75 (1962).

239. G. V. Samsonov, V. V. Vedeneeva, and Kim Do Chir, "Problems of Fermentation and Purification of Antibiotics" (in Russian), Trudy Lenkhimfarminstituta, Leningrad, p.81 (1962).

240. G. V. Samsonov, V. V. Vedeneeva, L. N. Zav'yalova, and T. A. Vikhoreva, "Problems of Fermentation and Purification of Antibiotics" (in Russian), Trudy Lenkhimfarminstituta, Leningrad, p.93 (1962).

241. G. V. Samsonov, A. A. Selezneva, and Van i Guan, "Problems of

Fermentation and Purification of Antibiotics" (in Russian), Trudy Lenkhimfarminstituta, Leningrad, p.101 (1962).

242. G. V. Samsonov, N. P. Kuznetsova, R. B. Ponomareva, V. S. Pirogov, A. A. Selezneva, and Van i Guan, Zh.Fiz.Khim., 37:280 (1963).

243. G. V. Samsonov, V. V. Vedeneeva, A. A. Selezneva, and E. E. Voikhanskaya, Zh.Fiz.Khim., 37:725 (1963).

244. A. A. Selezneva, V. V. Vedeneeva, and G. V. Samsonov, "Ion-Exchange Technology" (in Russian), Nauka, Moscow, p.177 (1965).

245. V. V. Vedeneeva and T. A. Vikhoreva, "Selective Ion-Exchange Sorption of Antibiotics" (in Russian), Trudy Lenkhimfarminstituta, Leningrad, p.85 (1968).

246. R. M. Bakaeva and G. V. Samsonov, "Selective Ion-Exchange Sorption of Antibiotics" (in Russian), Trudy Lenkhimfarminstituta, Leningrad, pp.63,69 (1968).

247. R. M. Bakaeva and G. V. Samsonov, Antibiotiki, No.8, p.737 (1968).

248. R. M. Bakaeva, L. A. Domracheva, and S. S. Urusova, Khim.-farm.Zh., No.7, p.28 (1969).

249. R. M. Bakaeva and G. V. Samsonov, Antibiotiki, No.11, p.985 (1970).

250. R. M. Bakaeva, Yu. N. Sorokina, O. P. Kolomeitsev, and G. V. Samsonov, Khim.-farm.Zh., No.3, p.24 (1973).

251. R. M. Bakaeva, N. V. Fadeeva, N. A. Dedyurina, and G. V. Samsonov, Khim.-farm.Zh., No.8, p.94 (1978).

252. E. E. Howe and I. Putter, U. S. Patent 2,541,240 (February 13, 1951).

253. F. C. Nachud and J. Schubert (eds.), "Ion-Exchange Technology," Academic Press, New York (1956).

254. G. V. Samsonov and S. E. Bresler, Kolloidn.Zh., 18:88, 155, 337 (1956).

255. G. V. Samsonov, Kolloidn.Zh., 18:592 (1956).

256. G. V. Samsonov, S. F. Lavrent'eva, and M. P. Shesterikova, Antibiotiki, No.2, p.32 (1957).

257. S. E. Bresler, G. V. Samsonov, A. A. Selezneva, R. L. Goglozha, B. F. Kravchenko, S. D. Shevelkin, L. M. Shuvalova, and M. P. Shesterikova, "Antibiotic Separation and Purification" (in Russian), Goskhimizdat, Leningrad, p.14 (1959).

258. G. V. Samsonov and Yu. B. Boltaks, "Antibiotic Separation and Purification" (in Russian), Goskhimizdat, Leningrad, p.35 (1959).

259. G. I. Kil'fin and G. V. Samsonov, "Selective Ion-Exchange Sorption of Antibiotics" (in Russian), Trudy Lenkhimfarminstituta, Leningrad, p.74 (1968).

260. G. I. Kil'fin, V. A. Pasechnik, and G. V. Samsonov, "Selective Ion-Exchange Sorption of Antibiotics" (in Russian), Trudy Lenkhimfarminstituta, Leningrad, p.113 (1968).

261. L. F. Yakhontova, B. P. Bruns, and Yu. S. Chekulaeva, "Ion-Exchange Sorbents in Industry" (in Russian), Izd.AN SSSR, p.203 (1963).

262. G. S. Libinson and M. D. Slugina, Zh.Fiz.Khim., 39:2813 (1965).

263. L. F. Yakhontova, E. M. Savitskaya, and B. P. Bruns, "Investigations of Ion-Exchange, Distribution, and Precipitation Chromatography" (in Russian), Izd.AN SSSR, p.3 (1959).

264. L. F. Yakhontova, "Investigations of Ion-Exchange Chromatography" (in Russian), Izd. AN SSSR, p.179 (1957).

265. L. F. Yakhontova, E. M. Savitskaya, and B. P. Bruns, Zh.Fiz.Khim., 33:15 (1959).

266. L. F. Yakhontova, E. S. Vaisberg, S. N. Kobzieva, N. L. Isaeva, and E. M. Savidkaya, Khim.-farm.Zh., 9:No.11, p.32 (1975).

267. É. S. Vaisberg, N. N. Shelenberg, E. M. Savitskaya, and G. S. Kolygina, "Ion-Exchange Materials in Science and Engineering" (in Russian), Nauka, Moscow, p.151 (1969).

268. K. K. Kalnin'sh, B. V. Moskvichev, L. V. Dmitrenko, B. G. Belen'kii, and G. V. Samsonov, Izv. AN SSSR, Seriya Khim., p.1897 (1965).

269. G. V. Samsonov and N. P. Kuznetsova, Kolloidn.Zh., 20:209 (1958).

270. J. L. Haynes, J.Colloid and Interface Sci., 26:No.1, p.255 (1968).

271. N. P. Kuznetsova, B. V. Moskvichev, and G. V. Samsonov, Izv.AN SSSR, Seriya Khim., p.578 (1964).

272. N. N. Nemtsova, V. A. Pasechnik, and G. V. Samsonov, Zh.Fiz.Khim., 47:2398 (1973).

273. V. A. Pasechnik, N. N. Nemtsova, and G. V. Samsonov, Zh.Fiz.Khim., 48:1993 (1974).

274. V. A. Pasechnik, N. N. Nemtsova, A. K. Terk, and G. V. Samsonov, Zh.Fiz.Khim., 50:2235 (1976).

275. M. Seno and T. Jambete, Bull.Chem.Soc.Jap., 33:1532 (1960).

276. M. Seno, Bull.Chem.Soc.Jap., 34:1021 (1961).

277. J. Faitelson, Biochem.and Biophys.Acta, 66:229 (1963).

278. J. Faitelson, J.Phys.Chem., 65:975 (1961).

279. P. S. Nys and E. M. Savitskaya, "Ion-Exchange Technology" (in Russian), Nauka, Moscow, p.130 (1965).

280. B. P. Bruns, "Ion-Exchange and Chromatography Theory" (in Russian), Nauka, Moscow, p.90 (1968).

281. E. M. Savitskaya, P. S. Nys, and M. S. Bulycheva, Khim.-farm.Zh., 7:32 (1969).

282. P. S. Nys and E. M. Savitskaya, "Ion Exchange and Chromatography" (in Russian), Izdat. Voronezhskogo Gosuniversiteta, Voronezh, p.41 (1971).

283. E. M. Savitskaya, P. S. Nys, and M. S. Bulycheva, "Ion Exchange and Chromatography" (in Russian), Izdat. Voronezhskogo Gosuniversiteta, Voronezh, p.77 (1971).

284. M. S. Bulycheva, P. S. Nys, and E. M. Savitskaya, Zh.Fiz.Khim., 44:3099 (1970).

285. P. S. Nys and E. M. Savitskaya, Dokl.AN SSSR, 176:873 (1967).

286. L. K. Shataeva, A. A. Selezneva, O. V. Orlievskaya, D. I. Ostrovskii, V. N. Korshunov, R. B. Ponomareva, and K. P. Papukova, "Synthesis, Structure, and Properties of Polymers" (in Russian), Nauka, Leningrad, p.225 (1970).

287. L. K. Shataeva, I. A. Chernova, G. V. Samsonov, G. D. Kobrinskii, V. D. Solov'ev, and I. V. Domaradskii, Voprosy Med.Khim., 24:No.4, p.569 (1978).

288. G. A. Babenko, A. A. Selezneva, O. V. Varnavskaya, and G. V. Samsonov, Mikrobiol.Prom., No.1, p.14 (1972).

289. E. N. Vorob'eva, L. K. Shataeva, E. N. Datutashvili, and G. V. Samsonov, Prikl.Biokhim. i Mikrobiol., 12:715 (1976).

290. L. K. Shataeva, I. A. Chernova, and G. V. Samsonov, Izv.AN SSSR, Seriya Khim., p.358 (1977).

291. L. K. Shataeva, I. A. Chernova, P. D. Kobrinskii, V. D. Solov'ev, I. V. Domaritskii, and G. V. Samsonov, Voprosy Med.Khim., 24:569 (1978).

292. L. K. Shataeva, O. V. Pisarev, and G. V. Samsonov, Vysokomol.Soed., 22b:502 (1980).

293. G. V. Samsonov and M. D. Fadeeva, Biokhimiya, 21:403 (1956).

294. L. V. Dmitrenko, D. I. Ostrovskii, and G. V. Samsonov, Vysokomol. Soed.,Seriya B, 14:859 (1972).

295. D. I. Ostrovskii, L. V. Dmitrenko, L. T. Leibson, and O. A. Yuaev, Khim.-farm.Zh., No.8, p.49 (1973).

296. D. I. Ostrovskii, L. V. Dmitrenko, N. D. Sidorova, L. M. Smirnova, O. V. Chaika, V. L. Konstantinov, and E. I. Lyustgarten, Khim.-farm.Zh., No.7, p.96 (1976).

297. V. S. Pirogov, V. A. Ozole, M. Kh. Rubene, and G. V. Samsonov, Izv.AN SSSR,Seriya Khim., p.181 (1980).

298. V. S. Pirogov, M. Kh. Rubene, and G. V. Samsonov, Zh.Prikl.Khim., 52:760 (1979).

299. V. S. Pirogov, L. V. Dmitrenko, and G. V. Samsonov, Prikl.Biokhim. i Mikrobiol., 3:363,628 (1967).

300. V. S. Pirogov, L. V. Dmitrenko, and G. V. Samsonov, "Ion-Exchange

Thermodynamics" (in Russian), Nauka i Tekhnika, Minsk, p.162 (1968).

301. N. Appelzweig, J.Am.Chem.Soc., 26:1061 (1944).

302. N. Appelzweig and S. E. Ronzone, Ind.Eng.Chem., 38:576 (1946).

303. F. G. Arraras, Chem.Abstr., 47:1865 (1953).

304. A. W. Kingsburg, A. B. Mindler, and M. E. Gilward, Chem.Eng.Prog., 44:497 (1948).

305. W. G. H. Edwards, "Chemistry and Industry," p.488 (1953).

306. N. A. Izmailov and S. Kh. Mushinskaya, Dokl.AN SSSR, 100:101 (1955).

307. Yu. V. Shostenko, Med.Prom.SSSR, No.2, p.35 (1955).

308. N. A. Izmailov and S. Kh. Mushinskaya, Zh.Fiz.Khim., 36:1210 (1962).

309. Yu. V. Shostenko, Yu. I. Ignatov, T. N. Gubina, L. T. Drabot, E. V. Vikherskaya, and M. A. Pinzhura, Khim.-farm.Zh., No.1, p.49 (1967).

310. N. G. Bozhko and Yu. V. Shostenko, Khim.-farm.Zh., No.12, p.31 (1971).

311. Yu. P. Temirov, Yu. V. Shostenko, and A. T. Shein, Khim.-farm.Zh., No.10, p.114 (1978).

312. A. T. Shein, S. Kh. Mushinskaya, Yu. P. Temirov, and Yu. V. Shostenko, Khim.-farm.Zh., 12:No.10, p.111 (1978).

313. G. N. Al'tshuler, M. A. Portnov, T. A. Kozulina, and L. A. Sapozhnikova, Khim.-farm.Zh., No.3, p.42 (1968).

314. G. N. Al'tshuler, Yu. S. Poradnova, and Yu. A. Makarov, Khim.-farm.Zh., No.7, p.38 (1969).

315. G. N. Al'tshuler, E. A. Savel'ev, and N. A. Shbovich, Khim.-farm.Zh., No.8, p.20 (1970).

316. G. N. Al'tshuler and L. A. Sapozhnikova, Khim.-farm.Zh., No.9, p.34 (1971).

317. G. N. Al'tshuler and E. A. Solov'ev, Khim.-farm.Zh., No.12, p.22 (1971).

318. G. N. Al'tshuler and E. A. Solov'ev, Zh.Fiz.Khim., 45:1301 (1971).

319. ibid., p.2270.

320. G. N. Al'tshuler and L. A. Sapozhnikova, Zh.Fiz.Khim., 45:2897 (1971).

321. G. N. Al'tshuler and A. I. Fomchenkova, Zh.Fiz.Khim., 46:2124 (1972).

322. G. V. Samsonov, Dokl.AN SSSR, 97:707 (1954).

323. J. Wilson, J.Am.Chem.Soc., 62:1583 (1940).

324. D. De Vault, J.Am.Chem.Soc., 65:532 (1942).

325. J. Weiss, J.Am.Chem.Soc., p.297 (1943).

326. G. V. Samsonov, "Chromatography," B. P. Nikol'skii (ed.) (in Russian), Izdat. Lengosuniversiteta, Leningrad, p.22 (1956).

327. N. S. Rakutina, G. E. El'kin, and G. V. Samsonov, Kolloidn.Zh., 41:83 (1979).

328. G. V. Samsonov and V. N. Orestova, Kolloidn.Zh., 19:615 (1957).

329. G. E. El'kin, S. F. Klikh, and G. V. Samsonov, Zh.Prikl.Khim., 33:1397 (1960).

330. G. V. Samsonov, G. E. El'kin, and A. I. Gitman, "Problems of Fermentation and Antibiotic Purification" (in Russian), Trudy Lenkhimfarminstituta, Leningrad, p.211 (1962).

331. G. V. Samsonov, A. P. Bashkovich, V. T. Gvozdeva, and L. A. Moiseenko, "Problems of Fermentation and Antibiotic Purification" (in Russian), Trudy Lenkhimfarminstituta, Leningrad, p.185 (1962).

332. G. V. Samsonov, L. M. Shuvalova, M. P. Shesterikova, S. F. Lavrent'eva, O. V. Maslennikova, A. A. Kononova, and V. V. Bokareva, Kolloidn.Zh., 18:474 (1956).

333. G. V. Samsonov, L. V. Dmitrenko, A. T. Sirota, A. D. Goryunkova, I. G. Morozova, S. F. Klikh, and M. P. Shesterikova, Antibiotiki, No.2, p.90 (1958).

334. G. V. Samsonov and A. P. Bashkovich, Med.Prom.SSSR, No.10, p.5 (1959).

335. G. V. Samsonov and A. A. Selezneva, "Antibiotic Separation and

Purification" (in Russian), Trudy Lenkhimfarminstituta, Leningrad, p.40 (1959).

336. G. V. Samsonov, S. F. Klikh, and M. P. Karpenko, "Antibiotic Separation and Purification" (in Russian), Trudy Lenkhimfarminstituta, Leningrad, p.48 (1959).

337. G. V. Samsonov, S. V. Kol'tsova, L. K. Shataeva, N. N. Kuznetsova, K. M. Rozhetskaya, I. K. Paradeeva, and Z. D. Fedorova, Avt.Svid. SSSR (USSR Patent) No.584034, Byull.Izobret., No.46, p.60 (1977).

338. G. É. El'kin, S. F. Klikh, and G. V. Samsonov, "Ion-Exchange Technology" (in Russian), Nauka, Moscow, p.146 (1964).

339. K. É. Papiel', Kh. Ya. Kal'yula, N. Yu. Sakaluskaite, E. V. Letunova, A. S. Tikhomirova, L. K. Shataeva, and G. V. Samsonov, Prikl. Biokhim.i Mikrobiol., 11:598 (1975).

340. G. É. El'kin, A. P. Bashkovich, and G. V. Samsonov, "Ion Exchange and Ionites" (in Russian), Nauka, Leningrad, p.156 (1970).

341. F. Helfferich, Bi-ionic potentials, Disc.Faraday Soc., 21:83 (1956).

342. R. Schlogl and F. Helfferich, Comment on the significance of diffusion potentials in ion-exchange kinetics, J.Chem.Phys., 26:5 (1957).

343. F. Helfferich and M. S. Plesset, Ion-exchange kinetics. A nonlinear diffusion problem, J.Chem.Phys., 28:418 (1950).

344. F. A. Closki and I. S. Dronoff, Ion-exchange kinetics. A comparison of models, Am.Inst.Chem.Eng., 9:426 (1963).

345. N. V. Bychkov, Yu. P. Znamenskii, and A. I. Kasperovich, "Studies of the Properties of Ion-Exchange Materials" (in Russian), Moscow, p.30 (1964).

346. Yu. P. Znamenskii, A. I. Kasperovich, and N. V. Bychkov, Zh.Fiz.Khim., 42:2017 (1968).

347. R. M. Barrer and R. F. Bartholomew, J.Phys.Chem.Solids, 24:309 (1963).

348. C. W. Kuo and M. M. David, Am.Inst.Chem.Eng., 9:365 (1963).

349. B. Hering and H. Bliss, Am.Inst.Chem.Eng., 9:495 (1963).

350. R. M. Barrer (ed.), "Diffusion in and Through Solids," CUP, Cambridge (1941).

351. G. E. Boyde, A. W. Adamson, and L. S. Myers, J.Am.Chem.Soc., 69:2836 (1947).

352. F. Helfferich, Disc.Faraday Soc., 21:83 (1956).

353. E. P. Cherneva, V. V. Nekrasov, and N. N. Tunitskii, Zh.Fiz.Khim., 30:2185 (1956).

354. G. Ya. Gerasimov, L. F. Yakhontova, and B. P. Bruns, Vysokomol.Soed., Seriya A, 2:864 (1960).

355. A. A. Selezneva, G. A. Babenko, and G. V. Samsonov, Kolloidn.Zh., 36:511 (1974).

356. C. G. Horvath, B. A. Preiss, and S. R. Lipsky, Anal.Chem., 39:1422 (1967).

357. I. I. Kirkland, J.Chromatogr.Sci., 7:361 (1969).

358. P. Ya. Polubarinova-Kochina, "Theory of the Motion of Subsurface Water" (in Russian), Moscow (1952).

359. G. M. Kondrat'ev, "Regular Thermal Regimes" (in Russian), Moscow (1954).

360. A. V. Lykov, "Theory of Thermal Conductivity" (in Russian), Moscow (1967).

361. N. N. Tunitskii and E. P. Cherneva, Zh.Fiz.Khim., 24:1350 (1950).

362. N. N. Tunitskii and I. M. Shenderovich, Zh.Fiz.Chem., 26:1425 (1952).

363. N. N. Tunitskii, E. P. Cherneva, and V. I. Andreev, Zh.Fiz.Khim., 28:2006 (1954).

364. Ya. V. Shevelev, Zh.Fiz.Khim., 31:960 (1957).

365. ibid., p.1210.

366. E. Kucera, J.Chromatogr., 19:237 (1965).

367. N. N. Tunitskii, "Diffusion and Random Processes" (in Russian), Novosibirsk (1970).

368. N. N. Tunitskii, V. A. Kaminskii, and S. F. Timashev, "Methods of

Physico-chemical Kinetics" (in Russian), Moscow (1972).

369. A. M. Voloshuk, P. P. Zolotarev, and V. I. Ulin, Izv.AN SSSR, Seriya Khim., No.6, p.1250 (1974).

370. P. P. Zolotarev and V. I. Ulin, Izv.AN SSSR,Seriya Khim., No.12, p.2829 (1974).

371. P. P. Zolotarev and M. M. Dubinin, Dokl.AN SSSR, 210:136 (1973).

372. P. P. Zolotarev and V. I. Ulin, Izv.AN SSSR,Seriya Khim., No.12, p.2858 (1974).

373. N. I. Nikolaev, P. P. Zolotarev, Yu. M. Popkov, and V. I. Ulin, "Theory and Practice of Sorption" (in Russian), Voronezh, p.12 (1981).

374. M. M. Senyavin, R. N. Rubinshtein, E. V. Venetsianov, N. K. Galkina, I. V. Komarova, and V. A. Nikashina, "Basic Calculation and Optimization for Ion Exchange" (in Russian), Moscow (1972).

375. I. A. Myasnikov and K. A. Gol'berg, Zh.Fiz.Khim., 27:1311 (1953).

376. I. B. Rosen, J.Chem.Phys., 20:387 (1952).

377. I. B. Rosen, Ind.Eng.Chem., 46:1590 (1954).

378. G. A. Aksel'rud, Inzh.Fiz.Zh., 11:No.1, p.93 (1965).

379. P. P. Zolotarev and L. V. Radushkevich, Izv.AN SSSR,Seriya Khim., No.8, p.1906 (1968).

380. P. P. Zolotarev, Izv.AN SSSR,Seriya Khim., No.10, p.2327 (1969).

381. P. P. Zolotarev, Izv.AN SSSR,Seriya Khim., No.8, p.1703 (1970).

382. G. É. El'kin, Yu. Ya. Lebedev, N. N. Momot, and G. V. Samsonov, "Short Reports of the Session on Mass Exchange in Solid-Liquid Systems" (in Russian), Tashkent, p.28 (1971).

383. G. É. El'kin, Yu. Ya. Lebedev, and G. V. Samsonov, Diffusion, in: "Kinetics and Dynamics of Physical Adsorption" (in Russian), Moscow, p.153 (1973)

384. G. V. Samsonov, G. É. El'kin, Yu. Ya. Lebedev, and N. N. Momot, "Ionites and Ion Exchange" (in Russian), Leningrad, p.98 (1975).

385. Yu. Ya. Lebedev, G. É. El'kin, N. N. Momot, and G. V. Samsonov, "Ion Exchange and Chromatography" (pt.1) (in Russian), Voronezh, p.49 (1971).

386. Yu. Ya. Lebedev and G. V. Samsonov, Zh.Fiz.Khim., 50:534 (1976).

387. P. P. Zolotarev, L. I. Kataeva, and V. I. Ulin, Izv.AN SSSR,Seriya Khim., No.12, p.2657 (1977).

388. G. É. El'kin and G. V. Samsonov, "Ionites and Ion Exchange" (in Russian), Nauka, Leningrad, p.102 (1975).

389. G. V. Samsonov, G. É. El'kin, O. P. Kolomeitsev, T. V. Atabekyan, and V. V. Azanova, "Ionites and Ion Exchange" (in Russian), Nauka, Leningrad, p.108 (1975).

390. ibid., p.110.

391. ibid., p.130.

392. G. A. Tishchenko, N. M. Chernysh, N. N. Kuznetsova, A. N. Libel', and G. V. Samsonov, Kolloidn.Zh., 40:571 (1978).

393. G. V. Samsonov, G. A. Tishchenko, Yu. Ya. Lebedev, N. N. Kuznetsova, and A. N. Libel', Kolloidn.Zh., 38:393 (1976).

394. G. V. Samsonov, G. É. El'kin, O. P. Kolomeitsev, and S. F. Belaya, Zh.Prikl.Khim., 51:805 (1978).

395. S. F. Belaya, I. M. Ogorodnova, O. P. Kolomeitsev, P. I. Zaitsev, and G. É. El'kin, Zh.Prikl.Khim., 51:1006 (1978).

396. M. A. Gavrikova, G. V. Samsonov, V. Ya. Vorob'eva, G. B. Zvyagintseva, and L. T. Belkina, "Ionites and Ion Exchange" (in Russian), Nauka, Leningrad, p.120 (1975).

Index

Hormone synthesis, 1
Hydration
  affecting ion exchange, 72-73
  of network polyelectrolytes, 29
Hydrodynamics, 127
Hydrogen bonds, 97
Hygromycin B, 123
Hypotension, removal of components
        causing, 54

Immunobiological enzymes, 1
Industrial scale separation, 5-6
Insulin
  diffusion of, 29, 33, 132
  in macroporous ionites, 48
  sorption of, 27, 104
  synthesis of, 1
Interchain contacts, in network
        polyelectrolytes, 40
Internal diffusion kinetics, 127-128
Interphase exchange kinetics, 125
Interphase mass exchange, 3
Intramolecular mobility, 30-31
  of polymer chains, 40
Ion diffusion, 22
Ion exchange
  capacity, 43
  equilibrium, 42-107
  isotherms, 54-55, 68, 109
  kinetics, 21
  selectivity of, 63-67
  sorption, 6
Ionites
  beads, 131-134
  bidispersed, 8, 137-144
  biporous, 130
  compaction, 82
  grinding, 62
  isoporous, 10
  macroporous, 10
  particle size, 84-87
  pellicular, 8
  polymerized, 97
  pressure, 55
  structure, 42-43, 72, 79
  surface-layer, 137-144
  swelling, 67
Ionogenic groups, 42, 59, 63, 84
  accessibility of, 62
Isoelectric points, of proteins,
        102
Isomerizing enzymes, 2
Isoporous ionites, 10
Isotherm
  ion-exchange, 68
  of dynamic ion exchange, 109
  of equivalent ion exchange, 54-59

Kanamycin, 97-98
  sorption of, 69
Kinetic curves, 21-22

Kinetic-dynamic analysis, 7-9
Kinetic methods of separating
        mixtures, 3-6
Kinetic permeability, 28-29
Kinetics, 124-144
  of ion exchange, 21, 84
KRS ionites, 90-91

Light scattering, in ionization
        studies, 30
Linear sorption isotherm, 128-130
Liquid chromatography, 4
Low-pressure chromatography, 137-144
Lyophilic desiccation, 13
Lysine, 138
Lysozyme
  diffusion of, 133
  sorption of, 105

Macronet, 12
  ionites, 132
Macropores, 11
Macroporous copolymers, 24
Macroporous ionites, 10
Mass exchange, interphase, 3
Mass transfer kinetics, 124
Matrix density, 10
Mercury porosimetry, 8, 13
Mesopores, 11
Methacrylic acid, 78
Methionine, 131-132
Microbeads, 130-131
Microbe synthesis, 68
Microglobules, 23-25
Microgranules, 36-39
  ground, 37-39
Micropores, 11, 24
Microstates, 79-83
Mineral ions, sorption of, 52
Mixed diffusion kinetics, 128
Mobility, of polymer chains, 30-31
Molecular biology, 2
Molecular genetics, 2
Molecular relaxation parameters,
        30-31
Molecular sorption, 6
Monoaminocarboxylic acids, as
        counter-ions, 59
Monovinyl monomers, 10
Morphine, 107
Morphocycline, 91
  sorption selectivity for, 75
Multi-interaction processes, in
        separation of mixtures, 5-6

Naphthalene residues, 93
Neomycin, 97-98, 140
  sorption of, 69
Network densities, 87
Network polyelectrolytes, 10-41
  conformation of, 30

161